Energy for Rural
and
Island Communities
IV

Energy for Rural and Island Communities IV

Proceedings of the Fourth International Conference held at Inverness, Scotland, 16–19 September 1985

Edited by

JOHN TWIDELL

IAN HOUNAM and CHRIS LEWIS

Energy Studies Unit, University of Strathclyde, Glasgow, Scotland

Administrator

FIONA JORGENSON

PERGAMON PRESS

OXFORD · NEW YORK · TORONTO · SYDNEY · FRANKFURT
TOKYO · SAO PAULO · BEIJING

U.K.	Pergamon Press, Headington Hill Hall, Oxford OX3 0BW, England
U.S.A.	Pergamon Press Inc., Maxwell House, Fairview Park, Elmsford, New York 10523, U.S.A.
CANADA	Pergamon Press Canada, Suite 104, 150 Consumers Road, Willowdale, Ontario M2J 1P9, Canada
AUSTRALIA	Pergamon Press (Aust.) Pty. Ltd., P.O. Box 544, Potts Point, N.S.W. 2011, Australia
FEDERAL REPUBLIC OF GERMANY	Pergamon Press GmbH, Hammerweg 6, D-6242 Kronberg, Federal Republic of Germany
JAPAN	Pergamon Press, 8th Floor, Matsuoka Central Building, 1-7-1 Nishishinjuku, Shinjuku-ku, Tokyo 160, Japan
BRAZIL	Pergamon Editora Ltda., Rua Eça de Queiros, 346, CEP 04011, São Paulo, Brazil
PEOPLE'S REPUBLIC OF CHINA	Pergamon Press, Qianmen Hotel, Beijing, People's Republic of China

Copyright © 1986 Pergamon Books Ltd.

All Rights Reserved. No part of this publication may be reproduced, stored in a retrieval system or transmitted in any form or by any means: electronic, electrostatic, magnetic tape, mechanical, photocopying, recording or otherwise, without permission in writing from the publishers.

First edition 1986

Library of Congress Cataloging in Publication Data

Energy for rural and island communities, IV.
Includes indexes.
1. Power resources--Congresses. 2. Renewable
energy sources--Congresses. I. Twidell, John.
II. Hounam, Ian. III. Lewis, Chris.
TJ163.15.E5334 1986 333.79'09173'4 86-2553

British Library Cataloguing in Publication Data

Energy for rural and island communities
IV : proceedings of the fourth international
conference held at Inverness, Scotland, September
16-19, 1985.
1. Power resources
I. Twidell, John II. Hounam, Ian III. Lewis,
Chris, *1948-*
333.79'11'091734 TJ163.2

ISBN 0-08-033423-7

Printed in Great Britain by A. Wheaton & Co. Ltd., Exeter

Preface

This fourth conference has demonstrated the international nature
of innovative energy systems.   Delegates from 30 countries arrived
at Inverness to consider the needs of rural and remote communities
where energy supply limits economic development.   It became quickly
obvious that problems and solutions are of equal concern to groups
in both the so-called developed world and the less developed
countries.

The theme of the Conference was 'integrated energy systems', and most
of the papers discussed the use of one or two renewable supplies in
combination with conventional systems.   It is important to realise that
the basic principles of alternative energy systems have now been
 established and that considerable experience has been accummulated
of individual working devices.   The challenge today is to blend the
best options at a particular site into an economically viable and
reliable supply for the community concerned.

Likewise it is important to stress the need for efficient supply
systems and efficiency of energy use.   Outstanding examples presented
at the Conference were two Norwegian wave power stations constructed
near Bergen.   Each station uses a combination of physical effects to
enhance wave energy capture and to decrease the unit cost of the power
produced.   Throughout the conference other examples appeared of such
opportunities to make 'clever' systems using the advantages of small
scale installations.   In general smaller scale allows, (i) 'hot spots'

of renewable sources to be utilised, (ii) integration of several supplies,
(iii) substitution of a high price or depleted conventional source by
new alternatives, and (iv) the use of modern control systems usually
based on microprocessor instrumentation.

Once again, this series of conferences was able to demonstrate the range
of new and alternative energy sources. Many of the associated technologies
are now completely accepted in appropriate circumstances, e.g. biogas
from farming and waste products, passive solar design for new buildings,
grid connected wind turbine generators. The challenge today is to integrate
such proven technologies with conventional methods and with each other to
provide economic supply systems.   Present at the conference were
representatives for the 13-European-country study for such integrated
systems for farms.   This programme is being supported by the Food
and Agricultural Organisation (F.A.O) and the Development Programme
(UNDP) of the United Nations. International cooperation of this
scale, and crossing East-West political boundaries is extremely welcome.

At the closing session of the conference, participants expressed the clear
wish to have greater South-North cooperation amongst themselves. As a
result we decided to begin an information bulletin for circulation to
those attending ERIC conferences.   The purpose is to give participants
the opportunity to contact each other directly for visits, the exchange of
equipment, information etc.   The bulletin will be produced from the
Energy Studies Unit.   There is also the desire to hold conferences in
other countries, and the organisers would be pleased to see this happen.
The next major conference in the series is planned for September 14th to 17th,
1987.

John Twidell, Ian Hounam, Chris Lewis and Fiona Jorgenson

Energy Studies Unit   Telephone 041-552-4400
University of Strathclyde   Telex: 77472-UNSLIB
Glasgow
Scotland

Sponsors

The Conference was sponsored by:

Bank of Scotland

British Council

British Petroleum

British Renewable Energy Forum

Chivas Brothers Ltd.

UK Dept. of Energy

Grampian Regional Council

Highland Regional Council

James Howden & Co.Ltd.

Royal Bank of Scotland

Scottish Branch of British Wind Energy Assoc.

Scottish Development Agency

Scottish Solar Energy Group

Western Isles Islands Council

University of Strathclyde

The Sponsors, whilst supporting the aims of the Conference, did not necessarily endorse the views expressed by the Contributors.

Contents

Contents

TOPIC A

Community Energy
Systems and Experience

An Energy Overview of Canada's Remote Communities with Reference to the Remote Community Demonstration Program

D. E. Rodger

Remote Community Demonstration Program, Department of Energy, Mines and Resources, Ottawa, Canada

ABSTRACT

This paper provides an energy orientation of Canada's 374 communities which are not served by grid electricity or natural gas distribution systems. It reviews current and projected energy demand in consideration of energy conservation and alternative supply opportunities. It outlines the Remote Community Demonstration Program, a national program designed to help remote communities and their energy supply agencies to identify, assess and implement such opportunities. Finally, the paper presents four approaches to integrated energy systems.

KEYWORDS

Remote communities; Canada; energy conservation; alternative energy supply; integrated energy systems.

THE COMMUNITIES

Currently, there are 374 communities in Canada considered remote in the sense that they are not served by grid electricity or natural gas distribution systems. While a small portion of these communities will be so served over the next decade, most have virtually no option for such connection. With a total population of less than 200,000, they comprise approximately 0.8% of the total population of Canada. In the second largest country (by area) in the world, the land area they cover is miniscule.

In their common features, they are very different from major Canadian population centers.

Size

With an average population of approximately 550, they may range from 10

households to 15,000 people. Family sizes are often larger than the
Canadian norm.

Location

Located in northern latitudes, they are geographically isolated from the
commercial and industrial centers of Canada and have severe access
constraints (often limited to plane, barge, or ship) with resultant high
transportation costs.

Climate

Climate is characterized by relatively long, cold (down to -60°C) winters
with short (often hot) summmers. Growing seasons are limited by short
frost-free periods. Snow and ice are normal winter conditions.
Permafrost, or permanently frozen ground, is a common building constraint
for foundations and infrastructure. Above the Arctic Circle, periods of
24 hour darkness occur in winter and 24 hour daylight occur in summer.

Culture

Most of the communities are characterized by native (Indian or Inuit)
population and culture. Religion is normally traditional native or
Christian.

Economy

Often traditional economies based on fishing, hunting and trapping
predominate. Some communities are governmental administrative centers.
Some are based on forestry, mining or commercial fishing. Some have
tourism as a base. Many feature local crafts and fabrication. Per
capita incomes are usually low relative to the rest of the country.

Level of Service

Health, education, infrastructure and other community services are often
of a lower level of service than the Canadian norm.

CURRENT REMOTE COMMUNITY ENERGY USE

Current energy use is characterized by a high dependence on petroleum
fuels for power generation and space heating. Petroleum resources are
depleting, vulnerable to world-scale cost fluctuations, subject to supply
disruptions internationally or in delivery to the communities.
Consequently, energy supply to the communities is of considerable concern.

Communities below the treeline are also very dependent on wood fuel for
space heating. While wood harvesting is often a source of employment,
for climatic reasons tree growth is very slow and wood use may not always
be a sustainable option. Other energy resources such as solar, wind,
hydro, geothermal, peat, etc. exist but are generally of restricted
applicability among communities.

Fuel and electricity costs are high, as much as 15 times the costs in
southern Canada due to transportation costs. A litre of fuel oil (air
supply) in Fort Severn, Ontario exceeds $1.60. Electricity costs in
communities where oil is flown in exceed 50¢/kWh. In the south this cost

A1.2

may be 4¢/kWh. These high costs are offset to the consumer by a variety of governmental subsidies. While easing the consumer energy bill, these subsidies also reduce the incentive of consumers to conserve energy.

Subsidies are a major barrier to the pursuit of energy conservation and alternative energy supply options.

Total remote community energy use is summarized as:

- total energy	35.6	PJ
- per capita energy use	181	GJ
- oil for electricity	9.15	PJ
- oil for space heat	12.09	PJ
- cost of oil ($millions)	214.5	

The reasons to pursue energy conservation and alternative energy supply opportunities are two-fold:

- for the communities potential benefits include: improved and more secure energy supply, improved opportunities for economic and social development, improved social conditions (e.g. quality of housing), and in some cases reduced energy prices.

- to the energy supply agencies benefits include: reduced energy costs, consumer subsidy reductions, and increased commercialization opportunities for Canadian contractors and manufacturers. Based on known viable off-oil options, the projected reduction for energy requirements is 25% and for oil requirements 38%. All communities have viable energy conservation opportunities and many have alternative supply opportunities. The response of the Canadian government to these opportunities has been the Remote Community Demonstration Program.

REMOTE COMMUNITY DEMONSTRATION PROGRAM

To help reduce dependence on oil, the federal Department of Energy, Mines and Resources has established the Remote Community Demonstration Program. This Program is designed to help remote communities and agencies responsible for their energy supply to identify, assess and implement appropriate energy alternatives and energy conservation opportunities. To do this, the Program offers advice and limited financial assistance for carrying out studies to explore these opportunities. The Program also contributes funds to a limited number of demonstration projects in selected remote communities across Canada. It then transfers the knowledge gained from funded studies and projects to all communities and agencies where this knowledge is needed.

The Program aims to help communities achieve greater energy self-reliance, lower long-term energy costs and, in some cases, increased levels of electrical service for home, community and local industrial needs.

Eligible communities

To be considered under the Program, communities must be permanent (or have at least five years of planned life remaining) and have no less than

ten occupied principal residences, and not be connected to the
electricity grid or natural gas sytsems.

Eligible applicants

.   eligible communities or groups of communities as represented by
    recognized local organizations (preferably local governments)

.   companies (e.g., forestry, mining) that maintain eligible communities
    as part of their operations

.   provincial or territorial utilities

.   federal, provincial and territorial departments and agencies

.   other agencies responsible for, or directly interested in, energy
    supply and conservation for remote communities.

Technologies and measures

In the now completed study phase, a wide variety of energy options were
studied, including but not limited to: power supply systems based on
wood, wind, hydro, solar, geothermal and tidal resources and energy
conservation measures ranging from more efficient use of diesel systems
to building retrofits and energy efficient new construction. These
studies took place in a diversity of community situations. Many of these
studies have led to implementation by communities or energy supply
agencies. The priorities which have been identified for follow-up
demonstrations are:

Residential space heating

This accounts for 36% of energy used in a typical remote community.
Approximately 50% of residential space heating is by oil. Even in homes
using wood heat there are opportunities to use the fuel more
efficiently. Projects will address:

1. Improving the energy efficiency of new houses through improved design
    and construction techniques.

2. Improving the energy efficiency of existing homes through energy
    conservation measures and more efficient energy consuming equipment.

3. Substituting wood for oil as the major fuel for space heating and
    encouraging the safe, efficient use of wood.

Non-residential space heating

Non-residential buildings which account for 19% of energy consumption are
often high energy users. Projects will include:

4. Improving energy efficiency of institutional buildings through energy
    conservation measures and more efficient energy-consuming equipment.

5. Substituting wood for oil as the major fuel for space heating
    institutional buildings.

A1.4

Oil consumption in non-residential buildings may be reduced by linking buildings into a district heating system. Projects will include:

6. Wood fired district heating systems

7. District heating systems using waste heat from diesel generators.

Power generation

In most remote communities power is produced using diesel generators. Power generation accounts for 28% of remote community energy consumption. Projects to increase the efficiency of diesel use will include:

8. Load management and utilisation of waste heat.

Projects to encourage the substitution of oil for generating electricity will include:

9. Small hydro projects

10. Wood-fired generating systems

11. Wind turbines

General energy awareness

A significant problem is lack of awareness and lack of communication about energy issues. Activities will include:

12. Community energy planning activities

13. Conferences to bring together the interested parties to share ideas

14. Information transfer activities that are not project specific

INTEGRATED ENERGY SYSTEMS

The Program has applied the term "integrated energy sysems" in four ways.

Community - agency integration

Usually, it is a federal or provincial governmental agency which is responsible for energy supply and energy costs in a community. Often more than one agency may be involved.   On the other hand, the community as a whole and its residents are the consumers of energy. RCDP has been an effective vehicle to bring the implicated agencies together to cooperate and coordinate their activities for their own and the community's benefit and to bring the agencies and the communities together again for the purpose of consultation and coordination in respect of each other's distinct but related interests.

Energy planning and community planning integration

In the studies and demonstration it is supporting, RCDP is encouraging the planning of energy supply and conservation to not be isolated, uni-dimensional, single purpose activities. Energy is seen as a key

economic, social and physical influence on the community. Whether energy
provides new jobs or skills for the residents or whether the noisy new
diesel generator is located adjacent to houses can have a major positive
or negative effect on the community. Energy is not considered an end in
itself. As it is integrated into the community, it will contribute to or
detract from the community's larger objectives.

Energy conservation and alternative energy supply integration

RCDP encourages an integrated approach to meeting energy goals. It
serves little purpose to implement an alternative energy supply option if
wasteful energy demand continues. Similarly, energy conservation goals
can often be enhanced by implementing alternative supply options or
making existing supply systems more efficient (e.g. through waste heat
recovery from diesel power generation systems).

Technical integration

A variety of projects are being supported which feature technical
integration of energy systems. Examples include:

-   retrofitting diesel power generators with waste heat recovery or
    cyclical charge systems;

-   supplementing diesel power generators with add-on wind turbines to
    conserve fuel;

-   the use of combination wood/oil furnaces or boilers for space heating
    in houses and institutional buildings;

-   the use of cogeneration systems (based on a variety of fuels) to
    create both heat and power;

-   the development of hydro projects to provide both power and domestic
    water supply to the community.

The Fotavoltaic Project — a 50 kWp Photovoltaic Array on Fota Island

S. McCarthy and G. T. Wrixon

National Microelectronics Research Centre, University College, Cork, Ireland

ABSTRACT

This paper describes the Photovoltaic system which is used to supply electrical energy to a 250 head dairy farm on Fota Island in Cork, Ireland. Also described is an 80Wp system which supplies the complete electrical requirements of two elderly fishermen on a remote island located on the southwest coast of Ireland.

KEYWORDS

Photovoltaic system; Fotavoltaic Project; dairy farm; EEC Photovoltaic Program; remote island.

INTRODUCTION

The Fotavoltaic Project (Wrixon, 1983) is the name given to the 50kWp photovoltaic plant sited at Fota Island in Cork Harbour, Ireland, which supplies electrical power to a dairy farm. It is one of fifteen pilot projects ranging in size from 30kWp to 300kWp which have been built throughout Europe under the auspices of the Solar Energy R & D Programme of the EEC. The aim of this program is to gain experience at European Latitudes and with European climates of the effectiveness of medium sized photovoltaic systems to do useful work.

Powering a dairy farm using a solar generator was selected as the Irish dimension of this program because dairy farming is an important Irish industry and the energy demand of the dairy farm has the same seasonal variation as the output of a solar generator. During the summer months peak electrical demand occurs on the dairy farm due to increased milk yield and during winter months electrical demand decreases due to decreased milk yield and due to calving. Peak load demand of 115kWh/day exists during the summer months and this decreases to less than 5kWh/day in mid-winter. Solar generator output during the summer months exceeds 200kWh/day and decreases to 30kWh/day in mid-winter.

A2.1

Fig. 1   The Fotavoltaic Project

The system began operating in June 1983 and data recording began in January 1984. The first year involved gaining experience with the system's hardware, improving the software and improving the system's protection. During the peak milking period between March and August 1984, the PV array generated 20.52MWh, 4.643MWh was supplied to the dairy farm and 9.676MWh was delivered to the utility grid. Some energy was lost due to non-optimised software control and hardware faults. At present the system operates automatically and data are being continually recorded and analysed. The VAX 11/780 computer at the National Microelectronics Research Centre (20km from Fota) displays the PV system's operation on a graphics terminal using data received from Fota via a telephone link.

TECHNICAL DESCRIPTION OF PLANT

The main aims of this design were

- to supply the dairy farm loads independent of the utility grid.
- to deliver excess energy to the utility grid.
- to operate automatically.

To implement these aims the following design was adopted. It consists of a 50kWp photovoltaic array, a 600Ah lead-acid battery, three 10kVA self-commutated inverters which supply the dairy farm loads and a 50kVA line-commutated inverter which delivers excess energy to the utility grid. The complete system is controlled automatically by a microcomputer.

Fig. 2   Section of Battery Room

Solar Array

The 50kWp array consists of 2775 solar modules mounted on the roof of a south facing building and inclined at 45° to the horizontal. The modules are mounted on a galvanised steel structure in one plane as shown in Fig. 1.   The modules (type PQ 10/20/02) are supplied by AEG-Telefunken and are rated at 19.2W at STC (1000 W/m^2, 25°C).

The maximum recorded output from the array was 47kWp. The maximum energy generated over a one hour period was 37kWh.   Losses in energy were due to the increased module temperature and also the angle of incidence between the sun and the plane of the array.   The maximum ambient temperature recorded was 22°C and the corresponding module temperature was 50°C.   The glass covers of seventeen modules were damaged since installation.   Fifteen were damaged by overtightening and two were damaged by irregularities on the metal frame. Since these modules were installed, AEG-Telefunken have redesigned the module frame to avoid damage due to overtightening.   The new modules have special brackets for mounting.

Power Conditioning and Batteries

Power conditioning is performed using two types of inverters. Self-commutated inverters are used to supply the three phase AC loads and a line-commutated inverter is used to deliver excess energy from the system to the utility grid.

Three 10kVA self-commutated inverters known as 'solarverters' accept DC
inputs of 245V-315V and provide an output of 380V three phase AC. Switching
transistors are used to provide an output waveform thus eliminating the
quenching circuits required for thyristor operated inverters. The solarverters operate at 95% efficiency under full-load and are protected by an
immediate shut-down when a short circuit or overload occur.

A 50kVA line commutated inverter known as the 'mini-semi' is a thyristor
operated unit and the 50Hz signal is obtained from the utility grid.
Consequently, in the absence of the grid the inverter does not operate. This
unit delivers excess energy to the utility grid and it may also be used to
recharge the batteries from the utility grid.

Lead acid batteries supplied by Varta are used and each battery cell has a
nominal voltage of 2V and a capacity of 300Ah. One hundred and thirty four
battery cells are connected in series to form a battery group. Battery
charging using the solar array is regulated by the microcomputer and the
charging current is maintained between 50A and 65A by switching the sub-arrays
to and from the battery. Hydrogen and oxygen are recombined by a catalyst in
the recombinator thus returning the water to the electrolyte and minimising
battery maintenance. The batteries with recombinators are shown in Fig. 2.

Power Management and Control

The system is controlled using a Motorola 6800 based system (McCarthy, 1983).
The 48K of software, which was written in assembly language, is stored in
EPROMS in the microcomputer. The energy flow within the PV system is
controlled using a switching strategy i.e. the configuration of switches which
interconnect the solargenerator, battery, loads and grid is determined by the
solargenerator output, battery state of charge and load demand. Data recording
began in January 1984 and these data have been used as a dynamic input to
improve the system's performance. A computer model (McCarthy, 1985) is also
being developed and this will be used to optimise the system's performance,
estimate degradation in the array and battery, and dvvelop design rules for
future systems.

As well as controlling the system and recording data the micro-computer
monitors data on a computer terminal. This data is also sent to the VAX
11/780 system at the NMRC (20km from Fota) where the systems configuration
and energy flow are displayed on a graphics terminal.

MAINTENANCE

Presently, the system operates automatically and the solar building is
unmanned. The system is checked each day by personnel at the NMRC using the
remote telephone link. The system is visited weekly and all hardware is
visually inspected. A regular maintenance (every 3 months) involves:

- testing all subarrays
- examination of batteries (voltage, specific gravity, temperature)
- examination of all measuring equipment.

No module cleaning is necessary due to the abundance of rainfall in the region

BEGINIS ISLAND PROJECT

Beginis Island is located on the south west coast of Ireland and its
inhabitants are two elderly fishermen - Jim and Mike Casey. There is no
electricity supply on the island but during the installation of the
Fotavoltaic Project a simple photovoltaic system was installed. It consists
of four 19.2Wp modules, two car batteries, a diode and a voltmeter and it is
used to supply power to: a portable television, two 15W (12V) fluorescent
lights and a VHF radio.

Maintenance of the system is performed by the fishermen. The modules are
cleaned regularly and the battery water level is kept constant with distilled
water. The operating time of the television, lights and radio is adjusted to
match the output of modules. Deepdischarge of the battery is prevented by
turning off all loads when the voltmeter indicates 11V.

This photovoltaic system is simple and reliable and since its installation in
May 1983 it has been 100% maintenance free.

The operation and maintenance is now part of the daily routine of the two men.

CONCLUSION

The main aim of the Fotavoltaic Project is to examine the performance,
reliability and efficiency of a Photovoltaic system at Northern European
Latitudes.

The main conclusion from this project is that it is possible to use
Photovoltaics as an energy source at Northern European Latitudes in a
situation where the load has the same seasonal variation as the insolation.

For large systems (kWp) the use of microcomputer control is essential for the
following reasons:

- Using software the systems performance can be continually optimised to
  maximise the use of the solar energy generated.
- This control may be tailored to specific local climates.
- Using software, hardware changes are minimised resulting in reduced costs
  and minimising system shutdown.

For remote applications the systems' design should be:

- Simple
- Reliable
- Easy to maintain.

For remote communities, it is important that the Photovoltaic system is an
integrated part of the community and that its maintenance and operation
become part of the daily routine of the community.

REFERENCES

McCarthy, S. and Wrixon, G.T. (1983). Development of a computer for the
    control of a photovoltaic power station. In W. Palz (Ed.), Photovoltaic
    Power Generation, Series C., Vol. 3, Reidel Dordrecht. pp. 230-235.

14   S. McCarthy and G. T. Wrixon

McCarthy, S. and Wrixon, G.T. (1985). The development of a computer model
    to optimise the performance of a 50kWp photovoltaic system. In W. Palz
    (Ed.), Photovoltaic Power Generation, Series C, Vol. 5, Reidel, Dordrecht.
Wrixon, G.T. (1983). The Fotavoltaic Project: A 50kWp photovoltaic system
    to power a dairy farm on Fota Island, Cork, Ireland. In W. Palz (Ed.),
    Photovoltaic Power Generation, Series C, Vol. 4, Reidel, Dordrecht,
    pp. 153-159.

Energy and Village Development: the Basaisa Experience

S. Arafa

Science Department, The American University in Cairo,
113 Kasr Aini Street, Cairo, Egypt

ABSTRACT

The field experiences of introducing biogas plants and bicycles to the Egyptian village of Basaisa are reported. An appropriate mixture of mutually supporting technical, economic, institutional and developmental approaches was found essential. A community-based organization(COMMUNITY COOPERATIVE FOR DEVELOPMENT) ; was developed at Basaisa to manage and coordinate the different activities at the village level and to disseminate the knowledge and experience gained to other villages in the area.

KEYWORDS

Rural Energy; Renewable Energy Resources; Rural Community Development; Bicycles; Biogas; Cooperatives; Technology Transfer.

INTRODUCTION

There is no simple answer to the energy dilemma of the millions of small villages of the poor developing nations. Among the technical approaches which were proposed to ameliorate the present situation is the extensive use of renewable energy resources. This has stimulated global interest in appropriate technologies and has attracted widespread attention to the problems of Energy for rural communities as well as to the more general problem of Energy and community Development.(National Academy of Sciences,1976; International Energy Agency, 1978; UNU-UNESCO,1978).

In view of the fact that a community's welfare is some function of economic, social, technological and environmental factors, the criteria to be applied in evaluating viable energy supply systems when welfare maximization is assumed to be a basic goal in planning, design and operation and development of human communities should refer to the efficient use of the available productive resources of these communities (including their people) on the one hand and to the attainment of particular social, economic and environmental objectives, on the other.

Basaisa is a small satellite village with a total population of 318(Jan.1983

A3.1

survey) which lies in the heart of the Nile Delta at a distance of about
100 km North-East of Cairo in Al-Sharkiya Governorate of Egypt. Like the
nearly 30,000 other small villages in Egypt, Basaisa at the start of the
project(1978) had no direct access to state or regional services such as
Public Transportation, Electricity, Clean Water, Health Clinic,Youth Clubs
Draining system, Tel-Communication, Banks, Cooperative Societies, Schools,
-------etc. Obviously these facilities and organisations are becoming more
necessary for the development of rural communities. All of them are linked
directly or indirectly to issues of energy resources and use.

The Basaisa Integrated Field Project is concerned with exploring the possibilities, relevance and appropriatness of utilizing natural local resources
to meet the human needs of small rural communities. The project was based
from its inception on two fundamental premises:-(1) People's participation and
active involvement in whatever is going on in their community; and (2)Appr-
opriate utilization of local natural resources including renewable energy
resources for village development. The ultimate goals of the project are:
increasing the income and production of the village families; improving the
living conditions of the rural poor; integrating these activities in a comprehensive scheme for the development of Egyptian villages.

Several New and renewable energy systems as well as new economic and social
systems were introduced, field tested, and developed during the past 6 years
in the village of Basaisa. The results of the Solar powered micro-irrigation
pumps and the communal solar photovoltaic systems tested in Basaisa were reported at previous conferences on energy for rural and island communities
(Nelson,19 82; Arafa,1984).

This paper discusses the field experiences of Biogas plants and Bicycles
projects within the newly developed community- based organization: Community Cooperative for Development (CCD) at Basaisa. Some obstacles and constraints have been identified in the course of our work in the village. These
are discussed along with the future plans and activities of the Basaisa integrated field project and the potential of diffusion for area development.

PROFILE OF BASAISA

Basaisa is a small village community of 254 people (1974 survey) grouped into 63 families, living in 41 households in which the dominant mode of production is traditional agriculture. Tables 1-3 show some data on the setting
from surveys that were undertaken in 1974/75.

TABLE 1   Data on the Village Setting (1974)

Village Population	254	Water Sources(Indoor/Outdoor)	4/3/
- Males	146	Toilets(Indoor/Outdoor)	16/25
- Females	108	No. of Illiterates (M/F)	22/64
No. of Households	41	University Education (M/F)	1/None
- Mud bricks	40	No. of T.V. sets	None
- Red bricks	-	No. of Animals(Cattle)	161
- Mixed bricks	1	No. of male adults working in Agriculture	39
No. of Communal buildings	2	Distance to Mother village,km	3-4
No. of Families	63	Distance to Nearest town,km	8

TABLE 2   Land Ownership per Family in Basaisa -1975

Land Owned per Family(in Feddans)	No. of Families	Percentage
Landless	6	9.5
Less than one Feddan	24	38.1
1 ⎯	21	33.3
3⎯⎯	8	12.7
5 ⎯	2	3.2
7 ⎯	1	1.6
More than 10 Feddans	1	1.6
TOTAL	63	100

The field surveys showed that the annual cash income for 65% of the families
in Basaisa is less than 600 Egyptian pounds(Arafa and Hamid, 1982).

Basaisa represents then such a prevalent type of human communities in rural
Egypt possessing the least in terms of land and socio-economic resources
(Afifi, 1985) . These communities are characterized by a preponderance of
small fragmented land holdings and the majority of families are very poor
and a vicious cycle of poverty and dependence dominates (Nelson,1984;Arafa
and Nelson,1982). Agriculture residues represent about 98% of the total energy
consumed in Domestic Activities with 80% for cooking alone. In Basaisa agriculture residues are seldom bought and sold. Energy needs for irrigation and
other agricultural processes are nearly provided by human and animal power.
The village has good potential for solar energy utilization where the sunshine duration is about 4,000   hours/year and the global radiation on the
horizontal surface is 3 to 8.5 kWh/d. The renewable technologies could provide
energy for many productive activities as well as for basic services and creative activities to the thousands of the present small village communities
in the valley and Delta and to the new communities in the desert of Egypt.

PROJECT METHODOLOGY

The project can be described as "Participatory Discourse Oriented Action
Research" which is characterised by a continuous dialogue between actors and
researchers enlightening the actors as well as the researchers about the meaning of actions intended , and eventually resulting in increasing autonomy of
actors in relationship to researchers, and to an emancipation from question-
able and restraining beliefs in the inevitability of the given order of things.

From our initial encounter with Basaisa, we learned a mode of interaction
that became our basic pattern during every visit to the village since that
first Friday visit in 1974. Through the pattern of hospitality, where we met
inside the houses, small groups of people would gather around and would open
up discussions around issues of concern. This opened the way to talk about
new ideas,stimulated initiatives and built up the trust between the villagers
and the project team. From the households we moved to wherever people naturally
congregated(The Mosque and the Mandara) to open up further discussions and
lay the foundation for people's participation in local projects and for the
democratic process at the village level. Ideas are initiated through dialogue
and disc ussion in an atmosphere of egalitarian exchange in which the project
team members listen to and try to stimulate the villagers to "raise their
voice". Our role on-site is to help clarify the intentions of action by pos-
ing questions for the villagers to answer and to answer questions posed by the
villagers. In other words we invite the inhabitants to engage in critical
discourse to formalize a plan of action to find a solution.

The general methodology used in introducing new or improved technologies to
the village community stimulates local innovations and indigenous technologies, increases the awareness and strengthens local participation in the process
of development·

BICYCLES

The technology is simple and well known. It is also available in the local
market. Every body can feel and express the need and advantages of using such
technology. But our field surveys showed that the main constraint is the high
initial cost! A normal bicycle is now available at a price of 85 Egyptian
Pounds in the local market. The majority of rural people can not afford paying
such amount as a single cash payment. Furthermore, the majority have no access
to formal credit. Many families (as much as 80%) spend, on the average, about
L.E. 5-6 per month per family for their short distance transportation plus
few hours of walking per family per day(Arafa and Badei,1983).

At the start of the project there was only one bicycle and one motorcycle in
the village. On discussing the energy needs for transportation, bicycles were
found to be the appropriate option for short-distance transportation. The main
problem was to find a system by which a villager could get a bicycle easily
as a loan even with an interest and pay on installments   over a period of
about 1-2 years.

When this was expressed as a need by a large group of the village inhabitants
the following questions were raised:-(1) Who is going to be responsible for
the bicycles project? (2) From where can we get the funds needed?(3) What is
initial capital needed and how to use it as a revolving fund to achieve the
final goal(a bicycle for each household who needs it)? (4) Who would be eligible and on what terms is he going to pay back the loan? (5) What are the
legalities and limitations on the project?.

An educated young man(Ismail Hussein) from the village volunteered to take the
responsibility for the implementation of the project. A loan of the amount of
L.E. 1277 was used by the Community Cooperative for Development at Basaisa as
a revolving fund for this particular purpose. In less than 3 years the loan
was paid back with the accumulated profit   now part of the capital for the
second phase of the project (Arafa and Badei,1983).

The responsibility of the project was smoothly passed over to another young
man(Mohamed Hassan) who lives in Basaisa.

Table 3 shows the status of the bicycles project to-date.

TABLE 3 The status of the bicycles project to-date(July 26,1985)

Date of implementation	Sept. 1, 1981
Total No. of Beneficiaries	56 members
-No. from Basaisa village	41 members (73.2%)
-No. from other communities	15 members (26.8%)
No. of Students	12 members (21.4%)
No. of Farmers	13 members (23.2%)
No. of Employers	22 members (39.3%)
No. of others	9 members (16.1%)

The time and energy saved by cycling were immediately utilized in more productive and creative activities within the community. This saving also allowed
many members to participate more actively in communal activities like aware-

ness and literacy classes. Every household in Basaisa today has at least one
bicycle saving a minimum of L.E. 50 per year from its limited cash income.

It is interested to note that the time saved by cycling per family for a day
allowed many to use it in the available income-generating activities and earn
what is equivalent to the monthly payment for a bicycle.

BIOGAS

Biogas Systems in recent years have attracted considerable attention as a
promising approach to decentralized rural development (UNU-UNESCO,1978).
An excellent overview on Biogas as a fuel for Developing Countries is given
by DaSilva(1979).

In this case the technology was not known and general awareness and some training were needed. R&D programs in Egypt are under way at the National Research
Center (NRC) and at the Agriculture Research Center (ARC). The primary problems are adaptation of designs appropriate to Egyptian Conditions and technology transfer to the field with familiarization of Egyptian farmers with
the new technology.

Three biogas digesters were built of locally obtainable materials such as
bricks, concrete, clay pipes and metal tanks. Construction was undertaken by
local artisans after a one-day training course and under constant supervision
of a well trained local person. Plastic piping for the gas was with locally
manufactured burners,valves,lamps and fittings. Figure 1 shows a biogas plant
with modified Indian specifications and fig. 2 shows a biogas plant with modified Chinese specifications built at Basaisa in cooperation with ARC.

Fig. 1 one of the biogas plants constructed with modified Indian Specifications at Basaisa village.The
responsible villager is inspecting
the digester.

Fig. 2 A biogas plant constructed with
modified Chinese Specifications at
the integrated Rural Technology Center
at Basaisa.

The cost for a family plant(about L.E. 650-950) was found very high and nearly
impossible for many rural families to pay in cash as one instalment.   Furthermore, it is also difficult to ask a family to pay as instalments more than
what they might pay for using the highly subsidised commercial energy sources.

A3.5

Therfore, the project decided to subsidise the construction costs of family plants with L.E.40 per m^3 and the remaining cost is to be paid in instalments over a period of 10 years (Arafa, Bamzai and Ibrahim,1983). The project will guarantee the operation of the plant for the first year.The 3 plants constructed in Basaisa (April 1983) are in successful operation to date. The produced gas is used primarily for cooking and heating water.

COMMUNITY COOPERATIVE FOR DEVELOPMENT

Local organizations, whether governmental or non-governmental are an essential prerequisite to community-based initiatives and participation in local devlopment projects.Rather than creating more government structure under the local units to address basic community needs, we developed in consultation with the villagers an indigenous private voluntary organization(PVO) at the village level which could fill the gap between individuals and existing local units. This newly developed organization was called the Community Cooperative For Development(CCD) at Basaisa(Arafa and Hamid,1982).To date, there are 141 members in the CCD and nearly all inhabitants of Basaisa are involved one way or another in its activities.About 27% of the members are from other villages.

PROBLEMS ENCOUNTERED

Some obstacles and constraints have been identified in the course of our work in Basaisa over the past several years (Lumsdaine and Arafa,1982). The most glaring of these is one which almost all renewable energy technologies face: the high initial cost of the systems compared to conventional systems and low operating costs of the latter, due to heavy subsidisation of fuels. Other obstacles include:water cycle for irrigation,land use,very low cash income, high illiteracy rate especially among women,scarcity of skilled labor and absence of two-way communication.

The communal approach through the developed CCD was found to be an attractive solution.It generates funds that can be used for services and for subsidizing the cost of the new technologies appropriate for the community. The CCD also provides information,awareness and training as investment on human and local resources.

REFERENCES

Afifi,E.A(1985). M.Sc.Thesis. Anthropology depart. Institute of African
    Studies,Cairo Univ. Egypt.
Arafa,S and M.Hamid(1982). AUC-NSF-Basasia Project Report, No. 53.
Arafa,S.,and C.Nelson(1982). AUC-NSF-Basaisa Project Report, No. 06.
Arafa,S.,and H.Badei (1983). AUC-NSF-Basaisa Project Report, No. 19.
Arafa,S.,A.Bamzai, and A.Ibrahim (1983). AUC-NSF-Basaisa Project Report, No.08.
Arafa,S.(1984) Communal Solar PV. systems for rural areas.'In' J.Twidell(Ed.)
    Energy for Rural and Island Communities II, Pergamon Press.
DaSilva,E.J.(1979). ASSET, vol. 1, No. 11, pp 10-16.
International Energy Agency(1978). Energy and Community Development. First
    International Conf. , Athens, Greece, July 10-15.
Lumsdaine, E.,and S.Arafa(1982). Improving World Energy Production and
    Productivity. Clinard,English and Ballinger Publishing Co., Chapter 10.
National Academy of Sciences(1976).Energy for Rural Dev. Washington,D.C.
Nelson,C.,S.Arafa, and R.Pearson(1982)Energy for Rural and Island Communities
    II, J.Twidell(Ed.) Pergamon Press,pp 359-365.
Nelson,C(1984) Cairo Papers in Social Sciences, vol.7, No. 3, pp 45-56.
UNU-UNESCO Symposium (1978). Bioconversion of Organic Residues for Rural
    Communities. Guatemala City, Guatemala.
World Bank (1983). The Energy Transition in Developing Countries. Washington,
    D.C.

Renewable Energy at Bornholm — Self-sufficiency for an Island Community in Denmark

K. Jørgensen

Physics Laboratory III, The Technical University of
Denmark, Building 309C, DK-2800 Lyngby, Denmark

ABSTRACT

This paper describes a project - carried out by the County Council of Bornholm - aiming at a drastic reduction of the use of imported energy at the
island of Bornholm. An energy system for the year 2010 is proposed built
around so-called "LOCUS-systems", that is small cogeneration systems in combination with heat pumps, thermal storages, windmills, biogasplants etc. A
"long-term development plan 1985-2010" is sketched out to ensure a smooth
transition to the proposed energy system. Questions concerning the activating
of local producers and potential users of the energy equipment are given
special attention.

KEYWORDS

Energy planning; self-sufficiency; combined systems; cogeneration; local
production; organizing questions;

INTRODUCTION

The island of Bornholm (47,000 inhabitants, 590 km^2) lies in the Baltic Sea,
150 km east of the rest of Denmark. The regional authority is the County
Council of Bornholm and there are 5 municipalities at the island. The project
for renewable energy at Bornholm (the "RE-project") was initiated in 1983 by
the County Council with the following purposes:

- to reduce the dependence on imported energy through increased efficiency of
energy use and conversion and increased utilization of local resources.
- to secure that the regional economy benefits from the transformation process
in the form of increased local production and employment
- to encourage and support potential users and producers to participate.

A period of 2-3 decades is used as time-perspective. It is emphasized that no
part of the transformation from the present energy system to the one in the
future should rest on decisions that are un-attractive to the involved
parties.

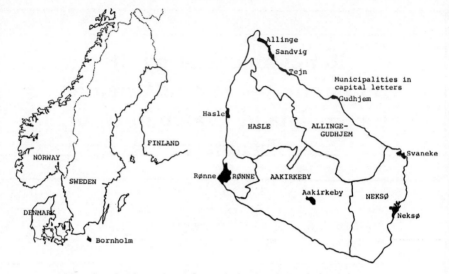

Fig. 1. The location of Bornholm in the Baltic Sea

The County Council has established an Energy Secretariat with a staff of four
persons to initiate the RE-project. Apart from the County Council, the Energy
Agency in Denmark and the municipalities among others participate in the
project, which has received economic   support from the Technology Council of
Denmark and the EEC.

                    THE ISLAND OF BORNHOLM

Bornholm is one of Denmark's regional development areas. The principal trades
are fishing, agriculture, forestry and tourism, while industry in the proper
sense is faintly developed. The biggest town is Rønne (15000 inhabitants).
The rest of the population divides equally between medium-size towns (1500-
4000 inhab.), small towns (200-1500 inhab.) and the open country.

The total energy turnover is roughly 5 PJ with more than 90% depending on
imported energy (fig. 2). Nearly all of the elctricity is supplied through a
cable from Sweden, mostly surplus electricity from the Swedish hydro and nuclear power, but also oil-/coalbased electricity from Sealand (Denmark)
transmitted to Bornholm via Sweden. There is, though, a power station in
Rønne based on coal and oil.

Heating is supplied almost exclusively through individual heating systems.
The only notable exception is a small district heating system based on wood
in the town of Neksø. Plans for district heating from the power station in
Rønne were rejected by the local council in Rønne. District heating systems
are meeting difficulties in Bornholm, particularly along the northern coast,
not seen elsewhere in Denmark, as there are rocks in the surface.

Bornholm has rather abundant local energy resources, most notably straw,
wood and wind resources. Fig. 3 shows the estimated quantities now and in
the future.

TJ/year	Heating	Transport	Process	Electricity	Total
Agriculture, Forestry 1)	20	25	150	65	260
Fishing 1)	15	795	60	40	910
Tourism 2)	110	20	0	35	165
Industry	15	105	240	115	475
Services & trade	400	120	10	120	650
Households	1465	300	40	175	1980
Professional transport	10	220	0	10	240
Total	2035	1585	500	560	4680

1). Including related industries.
2). Including hotels, restaurants, weekend cottages etc.
Ref.: Jørgensen, Lauge Pedersen & Obdrup, 1985.

Fig. 2. Energy balance for Bornholm in 1984.

TJ/year	1984	1995	2010
Straw	800	1330	1600
Wood	890	890	890
Biogas from manure	290	290	305
Municipal waste	175	215	285
Wastewater and sludge	15	15	15

Ref.: Jørgensen, Lauge Pedersen & Obdrup, 1985.

Fig. 3. Selected local energy resources at Bornholm by
utilization.

THE ENERGY SYSTEM AT BORNHOLM IN THE YEAR 2010

By the time of writing (Mid-August) the final dimensioning is being carried
out, and complete results of the calculations are about to appear. Therefore
only preliminary results are available at present. They show that Bornholm
can be made independent of imported energy within 2-3 decades, but since it
has been chosen to base the RE-project on rather conservative assumptions to
make it as realistic as possible, the proposed energy system for 2010 is not
totally based on local resources. The energy system is completed from known
technologies only, which is a very pessimistic assumption for a 25 year period.
Also it is assumed that the consumption of electricity, transport energy and
energy for industrial processes will remain largely constant throughout the
period. Most likely, though, it will drop in the last part of the period, as
new energy-efficient technologies penetrate the market. This would contribute
decisively towards the self-sufficiency perspective, since the production of
electricity causes the biggest problems. The heating demand is expected to
be reduced to about 70% of the present level due to insulation measures.

The "back-bone" of the energy system in 2010 are so-called "LOCUS-systems"
(Local Cogeneration Utility Systems) in most of the towns. The LOCUS_systems
(Illum, 1985, Lund&Rosager, 1985) are cogeneration systems consisting of four
components: Combustion engines, generators, heat pumps and thermal storages
(cf. fig. 4). The heat /power ratio of such systems is variable within a wide
range, and they have good regulating capabilities, which make them ideal for

combination with windmills and biogasplants. The version of the LOCUS-system proposed at Bornholm only has storage capacity for a few hours, but it may be extended to seasonal storage, when that is considered economical. In the small towns the LOCUS-systems are proposed supplied with biogas from manure and in the medium-size towns with gas from thermal gasification of wood and straw.

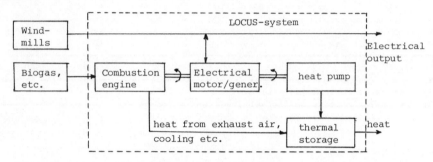

Fig. 4. Illustration of a LOCUS-system.

The LOCUS-systems contain windmills with an installed capacity of roughly 2-3 MW. In addition to that, windmills with a total capacity of 23 MW are established, including 5 MW in four windfarms. All windmills except those contained in the LOCUS-systems are placed at localities given first priority in a regional mapping of possible locations (Nybroe, 1985). Fig. 5 shows the proposed district heating systems and windmill locations.

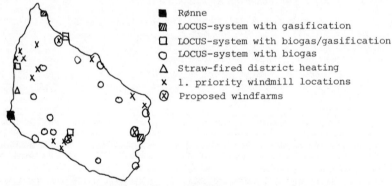

Fig. 5. Locations of district heating systems and windmills at Bornholm in the year 2010.

Outside the district heating areas the heating demand is met chiefly by individual straw- or woodfired boilers and biogasplants on some of the farms. Also 60,000 m^2 of solar collectors are expected to be established, mainly for production of hot water.

THE LONG-TERM DEVELOPMENT PLAN 1985-2010

A plan for the development of renewable energy at Bornholm in the period 1985 to 2010 ("The long-term development plan") is at present being prepared. The

purpose of this plan is to establish the best conditions for a smooth transformation from the present energy system to the one proposed for 2010, that
is to promote an even distribution of investments and work load over time and
to secure a coherent economy in the transition period.

The establishment of LOCUS-systems is divided into two stages, partly to
smooth out investments, partly to gain time for establishing adequate quali-
fications and experiences for the production of the systems at Bornholm. In
the first stage district heating systems based on comparatively cheap straw-
and wood-fired boilers are established. In the 1990's when these boilers are
worn out, they will be replaced by LOCUS-systems.

Outside the district heating areas the individual heating systems will at
first be converted to boilers burning wood-chips, as this is a known and
competitive technology. In the 1990's straw briquettes can be expected to
become competitive, leading to an ascending conversion to individual straw-
fired heating systems.

The total investments are in the region of 2-2.5 billion Danish kroners (150-
200 million English pounds).The economics of the plan are at present being
calculated and results will be available by the end of August. Preliminary
results indicate that the plan is favourable to the society in economic
terms if the positive effects on the employment are counted in. Without
subsidies the plan leaves the "typical consumer" with a deficit. On the other
hand the Treasury seems to profit considerably from the plan - enough to make
up the consumer's deficit and still keep a surplus. Apparently the grants
should be something like 50% of the investments or more, if the state were to
pay back all surplus to the consumers. Today the grants are typically 20-30%
in Denmark.

The development plan reduces the use of imported energy to around 20-25% of
the present level. The 75-80% is replaced by conservation measures (25%),
solar and wind (5%), increased efficiency of energy conversion (cogeneration,
20%) and utilization of local energy resources (25-30%).

INVOLVING USERS AND PRODUCERS

The purpose of the RE-project emphasizes the involvement of the users and
producers in the transformation process. This task has one of the staff at
the Energy Secretariat full-time employed.

To counter possible problems "manuals" for potential users and producers are
being prepared. To support local production of energy equipment s so-called
"master smith" strategy is used. This strategy was first used in Denmark
several years ago to promote local production of a 22 kW-windmill. Now it is
used to produce a 100 kW-windmill and in addition to that a "master smith"
boiler burning wood-chips is on its way. The centrepiece is a detailed
construction manual that the manufacturers wanting to produce the equipment
referred to have placed at their disposal. The first 100 kW-windmill has been
produced this way at Bornholm by 10-12 craftsmen collaborating on the project.
9 more windmills are to follow the first one to become the first windfarm at
the island.

Two questions are of crucial importance to the RE-project: Organizing the
distribution of local resources like straw, wood and manure and organizing
the district heating systems. The first of these problems concerns specifically the individual heating systems based on local resources, whereas the

26 K. Jørgensen

second concerns the problem of collecting different users with different
backgrounds around a collective energy supply. When it comes to practice the
"typical consumer" is of little relevance. This theme has been the topic of
several Danish studies (Nielsen&Tobiesen, 1983a, Nielsen&Tobiesen, 1983b,
Hvelplund, Rosager & Serup, 1983).

STARTING UP THE RE-PROJECT

A number of energy projects are being initiated   at the moment. In the town
of Lobbæk (400 inhab.) a straw-fired district heating system is in operation
and another is on its way in Klemensker (750 inhab.). A biogasplant at a big
farm supplying biogas for cogeneration at the farm and for the operation of
a tractor hast just been decided. The first windfarm is expected to be decided in the near future. In connection with that an electric car has been
made. The plans are that it should be loaded from the windfarm. A heat pump
based on seawater is planned in Svaneke, and a LOCUS-system has been proposed
for the town of Hasle, eventually with a supply of biogas for the buses   in
Rønne.

FINAL NOTICE

Most of the material mentioned in this paper will be available - in Danish -
by the end of September. This includes the long-term development plan. An
English translation of the long-term development plan is expected to appear
by the end of the year

REFERENCES

Hvelplund, F, F Rosager & K E Serup (1983). Hindringer for vedvarende energi,
    Aalborg.
Illum, K (1985). Preliminary report on integrated energy systems (in Danish),
    Aalborg.
Jørgensen, K, S Lauge Pedersen & J Obdrup (1985). Energibalancer for Bornholm
    1984-2010, Lyngby (forthcoming)
Lund H & F Rosager (1985). Systemundersøgelser. Udnyttelse af vedvarende
    energi på Bornholm, Aalborg.
Nielsen E M & F Tobiesen (1983a). Brugerorganisering, Århus.
Nielsen E M & F Tobiesen (1983b). Brugervejledning, Århus.
Nybroe, C (1985). Mølleparker på Bornholm. Regional placeringsundersøgelse.
    Metoder og resultater, Nimtofte.

Pay Attention to the Utilization of Renewable Energy in Chinese Islands

Gao Xiangfan

The Guangzhou Institute of Energy Conversion, Chinese
Academy of Sciences, PO Box 1254, Guangzhou, China

KEYWORDS

Chinese islands; renewable energy

I would like to present a report on the use of renewable energy in Chinese
islands.   Most of the islands lie far away from the mainland, in the East and
South of the China Sea.   The communication and transportation are not
convenient, and with no coal resources, there is always shortage of energy for
the communities.   Recently the economy has been developed very quickly, while
more and more electric power is required for industrial and domestic use.
Now energy has become an extremely important problem.   To develop the energy
in islands, the Chinese government encourages the use of diesel electric power,
as well as the use of natural resources, and pays attention to the use of
renewable energy.   The government advocates "depending on the local conditions,
using various sorts of energy to complement each other and comprehensive
utilization".   According to the distinguishing features and actual needs at
each island, there have developed solar energy, wind power, biomass energy
and ocean energy.   But there is a common weak point, that is unsteadiness
and interruption, so the renewable energy only can be used for saving coal
and other fossil fuel.

For example, the Zhousan Islands is the second largest group in China,
including about 600 small islands, the total area is 15,000 km^2, with
1,000,000 inhabitants.   The fishing and fish processing are well-developed,
the ship-building is being developed.   The income from the industries has
been more than that from the fishing.   Now the most pressing task is to
increase the energy.   A thermal power plant is being extended and a direct
electric current line is being built to connect with the grid on the mainland.
Renewable energy sources are also being encouraged.

The first is solar heating energy.   The total area of flat-plate solar
collectors is 3,000m^2, to heat water for communities.   This can save 370
tonnes of coal per year.   Two years ago, a solar desalination unit was built
at a port to change sea water into fresh drinkable water.

The second is wind power.   There are two large wind power generators, one of 18kW and another of 40kW, both of them have generated more than 90,000 kWh during these years.   In addition, there are some small wind mills for charging batteries and drawing water.

Methane-generating pits, which may be a speciality of China, are very useful in the islands.   A half century ago a methane-generating pit was built at a temple in the Puto Island which is a famous Buddhist spot.   Up to now more than 800 pits have been built in these islands; the total capacity is 26,000 m^3.   Crops stalks, animal soil, leaves and grass are all the raw and processed materials for the gas.   Most of the gas is used for heating water and cooking; some is used for generating electric power.   There are six small gas electric power stations built to provide power for lights, TV, electric fans and so on in the evening.

The utilization of ocean energy is being developed in the islands.   In 1970 a farmer who is an innovator got a success for researching the generation of electric power from tidal current.

A larger tidal power station has been built at a bay near the Zhousan Islands, whose total installed capacity is 3,000kW, with one unit of 500kW.   And some small tidal power stations have been built and some larger tidal power stations are being built at other islands.   A small fixed wave energy station is being designed at the South China Sea.   But the investment in ocean wave power stations is greatly more than that of the above mentioned renewable energy, so it is handled with great care by government.

The distinguishing features of the use of the renewable energy in Chinese islands are small-scale and comprehensive, so that it may be more practical and flexible, easy for construction, least for investment and convenient for management.   The communities in these islands are interested in using renewable energy, because it can support a part of electric power capacity. There is an old saying in China, "every little bit of collected hair goes toward making a fur".   That is what we have been doing to develop our use of renewable energy - small bits with better efficiency contribute to the economy of these islands.

Energy Resources, Supplies and Consumption: a Case Study of Some Rural and Remote Settlements in Nigeria

A. T. Atiku*, S. C. Bajpai*, C. E. C. Fernando** and A. T. Sulaiman*

*Sokoto Energy Research Centre, University of Sokoto, Sokoto, Nigeria
**Department of Chemistry, University of Sokoto, Sokoto, Nigeria

ABSTRACT

A survey of energy resources, supplies and consumption was conducted in three villages in the Sudan-Savanna region of Nigeria. Results of this survey indicate that people in these villages rely mainly on fuel wood, crop residues and animal wastes for their cooking and heating requirements. Supply of fossil fuels in these villages is highly irregular. There is no electricity supply to any of these villages. Human and animal power are greatly used for water collection, agriculture, transport and local crafts. A number of renewable energy systems on a family and community basis can make considerable improvements in the life-style of these dwellers.

KEYWORDS

Villages; Energy; Resources; Supplies; Consumption; Biomass; Wind; Solar; Renewable; Systems.

INTRODUCTION

This study was carried out in Sokoto State which is one of the four north-most states located in the Sudan-Savannah region of Nigeria. This region is presently affected by drought and desertification and has an annual rainfall of less than 700 mm. There are three seasons - the harmattan, hot, and rainy seasons. The harmattan season is characterised by dry dusty winds accompanied by substantial temperature falls, lasting from November to February. The rainy season is from July to September. The majority of the population in this region live in rural areas which most often have no electricity and hygienic water supplies.

Three villages in this region were chosen for study. These villages were visited several times to first get an idea of the life-styles of the people in relation to their socio-economic conditions and then a detailed questionnaire regarding their energy demands, supplies and consumption was used for the purpose of this study. This paper presents the results of the survey and advocates areas where renewable energy systems could be incorporated in order to improve the living conditions of the people.

GENERAL FEATURES OF THE VILLAGES

The three villages selected were Achida, Chacho, and Zumana. Achida is a
big village about 20km from Sokoto town. It has a population of about 8500
consisting of 1560 family units. 34.5% of the population are adult males,
27.7% adult females and 37.8% children. It lies on a major road, and is
the most developed of the three villages with facilities like a primary
school, secondary school, health clinic, and a big market. About 65% of the
adult male population are farmers   while the remaining are traders,
teachers, local craftsmen, etc.

Chacho.lies on a minor road and is about 30 km away from Sokoto town. It
has a population of about 3500 comprising 350 family units. The population
consists of 31.6% adult males, 23.7% adult females and 44.7% children. 75%
of the adult male population are farmers. It has a small market and a
primary school.

Zumana is in a remote location, about 12km from a small town 'Kware'. It is
the smallest and the least developed of the three villages and has a
population of 46 consisting of 5 family units. The population distribution
is 46.7% adult males, 30.0% adult females and 23.3% children. About half
of the adult male population are farmers. This figure is comparatively
lower due to small land-holdings per family unit, also many adults travel
frequently to nearby towns for employment.

There are no local industries in all these villages. Farming is carried
out during the rainy season only, and during the rest of the year, the
people are engaged in trading or crafts like the making of local fans, caps,
mats, thread, etc. Water supply is a major problem in all these villages.
Water is collected from distant wells and streams and in some cases purchased
from water tankers.

ENERGY RESOURCES, SUPPLIES AND CONSUMPTION

Biomass

Wood. There are a few shrubs and trees around these villages but there are
no organised fuelwood plantations. Wood is available in the local markets
in Achida and Chacho at about ₦1.00 per bundle (approximately 11kg). The
villagers in Zumana have to travel to a nearby village for their purchases,
and therefore manage with the wood they collect from nearby trees. For all
the three villages, firewood is scarce during the rainy season.

The main area where energy is consumed is cooking followed by space heating
and water heating during the harmattan. The villagers in Achida depend
solely on firewood for their heating purposes while those in Chacho supplement firewood with crop residues and in Zumana animal waste is also used.
In all these villages it is noted that the cooking system used - 3 stone
stove is very inefficient and there is considerable heat loss and waste of
firewood (Ahmed, and co-workers, 1985). The firewood consumption for the
3 villages is shown in Table 1 and Table 2, taking the calorific value of
fuelwood as 17,150 kJ/kg (Ahmed, and co-workers, 1985).

Forage and grasses. These grow widely and in some cases wildly during the
rainy season and are used extensively for feeding livestock.

Crop residues. Almost every family unit owns some land for farming. On an
average, a family unit in Achida owns about 6.5 acres, in Chacho 9 acres,
and in Zumana 4 acres. Millet, guinea corn and beans are grown during the
rainy season in Achida and Zumana while in Chacho groundnut is also grown.

Considerable amounts of crop residues are available after harvesting. The energy equivalent of the crop residues for the three villages is shown in Table 2, using the calorific value of crop residues as 11,600 kJ/kg (Energy/ Development International, 1984).

In Achida, the crop residues are used to feed livestock while in Chacho and Zumana they are also used to supplement the firewood for heating purposes.

Animal waste. Every family unit in these villages owns some livestock like cows, goats, sheep, donkeys and poultry. These are reared for investment purposes and in some cases as sources of animal power. The animals are fed on grass, crop residues and grains. On an average each family unit in Achida obtains about 7kg animal wastes daily while in Chacho it is about 15kg and Zumana 8kg. In Achida and Chacho this waste is used as manure in the farms, while in Zumana cowdung is also used to supplement the firewood. The energy resource from animal waste per family unit in a year is given in Table 2, taking the calorific value of animal wastes as 8,900 kJ/kg (Energy/ Development International, 1984).

Municipal waste. This consists of domestic waste and human waste. The people in these villages eat food prepared from millet or guinea corn and there is not much domestic waste available. This is either thrown into the nearby bushes or collected and burned. Human waste is disposed of in the pit-system.

Fossil Fuels

Kerosene. This is available in the local market in Achida and Chacho while those in Zumana purchase their kerosene from the nearby town. Kerosine is sold at about ₦1.00 per gallon. The villagers in Chacho and Zumana face the problem of irregular supply of this commodity.

Kerosene is used in lanterns or in other improvised systems for lighting. On an average, 11 hours of lighting is required per day for the 3 villages. Calorific value of kerosene is taken as 35,000 kJ/litre (UNDP/World Bank Report, 1983).

Gasoline and Diesel. The major transport facilities are public buses and vans. Villagers in Zumana do not have ready access to transport facilities due to their remote location and have to trek to the main road. Some villagers in Achida and Chacho possess motorcycles and bicycles. Travel is mainly to Sokoto town or nearby markets for trade, employment and sometimes to visit relatives. The average distance travelled per month per family unit is 440 km, 300 km and 250 km for Achida, Chacho and Zumana respectively. The fuel equivalent of these travels by public transport system is shown in Table 1. It is estimated that one person travelling 1 km by public transport requires about 0.017 litres of gasoline and the energy value of gasoline has been taken as 35,500 kJ/litre (UNDP/World Bank Report, 1983).

Cooking gas and coal. These are not available within the villages and are therefore not used.

Electricity

None of the villages has an electricity supply.

Animal Power

Some familes in all these villages own at least one donkey which is used for transport within the village for collection of water and also for going to

TABLE 1 Fuel Consumption Per Family Unit Per Month+

Purpose	ACHIDA		CHACHO		ZUMANA	
	Type of Fuel/Power	Fuel/Power Consumption	Type of Fuel/Power	Fuel/Power Consumption	Type of Fuel/Power	Fuel/Power Consumption
Cooking	Firewood	330kg	Firewood or Firewood/crop residues	480kg	Firewood/crop residues and Animal wastes	390kg (crop residues or equivalent 60kg.)
Water-heating	Firewood	135kg	Firewood or Firewood/crop residues or kerosene	150kg	Firewood/crop residues and Animal wastes	150kg (crop residues or equivalent)
Space-heating	Firewood	180kg	Firewood or Firewood/crop residues	240kg	Firewood/crop residues and Animal wastes	420kg (crop residues or equivalent)
Lighting	Kerosene	16.2 litres	Kerosene	12 litres	Kerosene	3.6 litres
Collection of water	Human Power Animal Power	150 Man hours 90 Animal Hours	Human Power Animal Power	180 Man hours 150 Animal Hours	Human Power Animal Power	270 Man hours 120 Animal hours
Transport	Diesel/gasoline Human Power	6.8 litres 60 Man hours	Diesel/gasoline Human Power	5.1 litres 82.5 Man hours	Diesel/gasoline Human Power	4.25 litres 75 Man hours
Local-Crafts	Human Power	150 Man hours	Human Power	150 Man hours	Human Power	150 Man hours
Agriculture	Human Power Animal Power	1365 Man Hours* 32 Animal Hours*	Human Power Animal Power	1935 Man Hours* 45 Animal Hours*	Human Power	860 Man hours* 20 Animal hours*

+ Except Agriculture
* Per season (June – September)

the farm. Animals are also used to carry the harvested crops and crop residues from the farm to the house. The animal power used for various activities is shown in Table 1 and Table 2 taking one animal-hour equal to 970 kJ (Action for Food Production, 1983).

Human Power

Considerable human effort is used in the daily collection of water from wells and streams. Human power is also the major input in farming for ploughing, sowing, weeding, fertilizer application and harvesting. The use of animal power or more mechanised systems is limited by the small extent of land owned by each family. The man-hours spent for various activities and the energy equivalent is shown in Tables 1 and 2, taking one man-hour as 840 kJ (Action for Food Production, 1983).

Wind Energy

Relatively strong winds are experienced during the harmattan season, the average wind speed being about 3.3 to 4.5m/s. Wind energy is only instinctively used for drying of crops and clothes.

Solar Energy

These villages receive abundant solar radiation, most of the year round. The average annual solar radiation is over 1800kWh/m^2 and an average of 10 sunshine hours per day. Solar energy is only used for drying of crops and clothes traditionally.

CONCLUSION

It is interesting to note that although Nigeria is an oil-exporting country, the commercial fuels like fossil fuels are not readily available in the rural areas and people still depend on traditional sources, particularly fuelwood. This trend need not be drastically changed due to the finiteness of fossil fuels. However, the felling of trees aggravates the existing desert encroachment in this region and could only be solved by massive fuelwood plantation programmes in the different villages. Also the firewood is wasted due to the inefficient cooking systems used and the change to more efficient cooking stoves and cooking pots is necessary.

There is a reasonable amount of animal waste in the villages which if used in biogas plants could provide fuel for cooking. At present this waste is used as manure, but the villagers need to be convinced that apart from the fuel gas, even better quality manure could be obtained by using these plants. Apart from biogas plants, firewood consumption could be significantly reduced by solar cookers which involve simple technology.

If the system of cooperative farming is encouraged, animal power and more mechanised systems could be used to reduce human power consumption in agriculture.

Water shortage is the most severe problem in these villages and should be given priority. This could be solved by windmills or photovoltaic pumping systems run on a community basis. It is also possible for the Local and State Governments to set up local industries run on a community basis and also photovoltaic television systems would be useful to promote literacy among the people.

REFERENCES
Action for Food Production, (1983). Decentralised Energy Planning for Food Production. New Delhi, India Page 22 - 23 .
Ahmed I, Bajpai S.C., Danshehu B.G. and Sulaiman A.T. (1985). Fabrication and Performance analysis of some traditional and Improved Woodstoves. Paper presented at the National Solar Energy Forum, Enugu, Nigeria.

A6.5

TABLE 2 Annual Energy Resources, Supplies and Consumption Per Family Unit (in Giga Joules)

Type of Energy	ACHIDA		CHACHO		ZUMANA	
	Resource/Supply	Consumption	Resource/Supply	Consumption	Resource/Supply	Consumption
Biomass						
Fuel Wood	Available in Market	84.12	Available in Market	118.85	Collected from Vicinity	80.53
Crop residues	14.44		30.07		13.92	
Animal waste	22.74		48.73		25.99	
Forage/grasses	During rainy season	Nil	During rainy season	Nil	During rainy season	Nil
Municipal waste	+	Nil	+	Nil	+	Nil
Fossil Fuels						
Kerosene	Available in Market	6.8	Available in Market	5.04	Not Available in village	1.51
Gasoline/diesel	Available in filling station	2.65	Not available in the village	1.99	Not Available in the village	1.66
Cooking gas	Not available	Nil	Not available	Nil	Not available	Nil
Coal	Not available	Nil	Not available	Nil	Not available	Nil
Electricity	No supply	Nil	No supply	Nil	No supply	Nil
Human Power	+	4.78	+	5.78	+	5.71
Animal Power	+	1.08	+	1.27	+	1.42
Solar Energy	+	+	+	+	+	+
Wind Energy	+	+	+	+	+	+

+ Difficult to compute

Note – Human and animal food, and fertilizers have not been included in this computation.

Energy/Development International (1984). Cookstove-News, Cottage Grove, U.S.A. Vol. 4, No. 1, Page 11.

UNDP/World Bank Report (1983). Nigeria: Issues and Options in the Energy Sector. Report No. 4440 – UNI, Page 135.

Island Alternative Energy — Matching Resources and Needs in Tropical Vanuatu

A. J. Garside

School of Mechanical Engineering, Cranfield Institute of
Technology, Cranfield, Bedford MK43 0AL, UK

ABSTRACT

Vanuatu exhibits different community needs through the islands, but common
needs relative to the individual. Natural resources are abundant and
economic exploitation is possible from wind and solar sources to provide
water, electricity, heat and cooling. Forest and dense bush, with a
management scheme to include charcoal manufacture has the potential to
provide energy to support main population centres. Other sources will
need future technology development to realise their potential. True needs
however must be identified.

KEYWORDS

Isolated islands, renewable energy, wind, solar, geothermal, biomass,
wave, town/village needs, electricity, potable water, coolers, heat.

INTRODUCTION

The Oceanic Republic of Vanuatu, Fig. 1, comprising 72 islands in 19
administrative groups and extending 1200km., contains few conventional
energy resources and limited though sufficient agricultural, marine and
mineral resources which could provide adequate convertible currency. The
country became independent from joint British and French condominium rule
in 1980. A result of condominium was the duplication of civil service and
education systems, and co-operative ventures, with a distinction between
Anglophones and Francophones having a common Melanesian background. Many
duplications have been removed and replaced by structures reflecting a Ni-
-Vanuatu outlook.

The economy is agriculture based, revenue earners being copra, fish and
coconut oil, with contributions from beef, cocoa, timber and coffee.
Imports are average for the region and include a significant oil fuel and
gas content. Proposals for the introduction of indigenous energy sources
have been presented to Government, (CPO 1980/1) and initial emphases have
been placed on plantation upgrading, livestock and forestry. The preliminary demonstrations using natural resources for power generation and heat

A7.1

have been completed, particularly through the
Rural Vocational Training Centre on Tanna
(Calvert 1980).

During a study tour, (Garside 1981), which was
confined to the southern and central islands,
various resources were investigated, in particular the wind resource, now looking the most
promising of the current renewable sources
(Bevan 1985), being exploited significantly in
the USA, Denmark and other European countries.

THE NEEDS

The Republic is well endowed with foodstuffs for
internal consumption and is able to grow
successfully non-indigenous crops. Livestock
farming has been extended from pig rearing,
(pigs are a symbol of status in village life) to
cattle and poultry rearing. A reliable water
supply is however essential for successful
ventures. Fishing has the potential for being
expanded for the local market and to enlarge the
export market; alternatives to fuel oil for
powering fishing vessels and cold stores would
be a great benefit.

Distribution of produce throughout the Republic
is a problem, especially to give access to the
thriving markets at Port Vila and Luganville.

Fig.1.   Vanuatu

Reliable and economic sea transport would improve communications and
increase the range of products available at the markets. Air Melanesae
provides a regular service through the group and carries goods and
livestock at a cost. These methods use imported fuel.

For sustained expansion in the agricultural and fishing sector a reliable
energy source, (currently fuel oil), is required to power equipment and
generation facilities for subsequent processing, freezing and
distribution.

Villagers need drinking water and some areas visited had barely adequate
resources in the wet season, necessitating removal to lower springs (more
brackish) in the dry season. Within a village there are central meeting
areas and with sundown occurring early, artificial light is desirable for
community activities. Pressure oil lamps, hurricane lamps or gas lamps
currently meet some of this need.

Each island or group has some level of health facility, from aid posts to
fullscale hospitals on the two main islands, Efate and Santo. The
tropical climate demands cooled storage to preserve medicines. (During
the visit a domestic refrigerator was encountered storing foodstuffs along
with drugs). Store cooling cycles may be limited since diesels are run
only to give power for short periods. Facilities available span diesel
powered electrical generation at large centres, to a limited clean water
supply at some aid posts.

Schooling throughout the Republic, being rationalised from the dual system has the needs of potable water, sanitation and for the future, means of teaching by radio or video. Schools on the main islands have either utility power or separate diesel supplies. Most regional schools are devoid of power.

The 300 cooperatives throughout the islands although not equally buoyant, have a background of operating shared facilities,(stores, a savings bank and abattoirs, etc.). The funding, operation and management of solar driers, radio communications and perhaps a centralised battery charging facility should be achievable.

Shaft power is used for milling of coconuts to provide oil, sawing and electrical power generation. Supplements to the diesel/petrol powered units may provide improved economics for the operations.

THE RESOURCES

A survey of the mineral resources of the group revealed only a little lignite and possible oil bearing strata under the surrounding sea area. Other mineral resources which could provide convertible currency are not extensive, and would be more economically exploited elsewhere. (Manganese mining has been terminated (1980)). There are materials suitable for building industry use, thus import substitution is possible.

Natural (renewable) resources are extensive, being available on land, at sea and from the sun. Land resources include geothermal sites, large areas of forest and bush whose wood and vegetation could be converted to provide gases, alcohols, charcoal and plant oils and some significant river systems which show potential for hydro electric generation up to 1.5MW. The direct solar resource has been used for drying, and with current technology, for water heating, and photovoltaic systems. Indirectly the sun provides the wind resource, which has been used for sail transport and could be used for electrical generation, water pumping and desalination. The ubiquitous Pacific swell although without the energy seen on UK coasts, may be amenable for conversion by emerging techniques. Use of the temperature gradients in the nearby trenches is possible but this will not be an early option. The sea depths are below 3000m. within the northern island group and in excess of 7000m., 100km. off western coasts of the southern group.

Draft and pack animals are not in use in Vanuatu and little can be done to introduce their usage, since Ni-Vanuatu are not accustomed to their use.

At any one village site it is unlikely that more than one resource could be exploited and a single source may be adequate for local needs. Exceptions are the large towns of Port Vila and Lurganville, which could be served by hydro-electric schemes and arrays of wind turbine generators.

SOME ASSESSMENTS

The Environment. A data bank for natural resources needs to draw from a long period of measurements which has been fairly rigorous. For Vanuatu the solar record ought to be accurate, but wind records give cause for concern.

The Met. Office site at Port Vila is surrounded by trees, at the edge of the town; Port Vila is on the lee side of the island. Data are received

from manned stations along the island group to provide comprehensive
forecasting, but stations at two outlying sites visited were sheltered,
one adjacent to an airstrip in the middle of the bush, the other having a
clear aspect to seaward only.

Wave energy potential has not been locally evaluated and the main
population centres are not well sited for transfer of generated power.
There are marked wet and dry seasons and rainfall records are extensive
for hydro resource evaluation.   No detailed studies of hydroelectric
generation have been completed but two large and several small scale
applications have been discussed.

The Flora. A general study has been made of the forest and bush resource
throughout the islands (CPO, 1980) and proposals made for improving the
quality and monetary value of the resource. Forest (dark bush) covers in
excess of 8×10^5 ha. of the islands (over 50%) having an energy potential
of 300×10^6 GJ. For comparison, in terms of indigenous needs, this is
equivalent to the UK coal resource.

There is potential for producing liquid fuels from plants. Manioc grows
locally and is a source of starch for alcohols. Many of the easily grown
plants have a high oil content, coconuts, peanuts and sunflowers for
example. A local tree fruited with the "candle nut", a natural night
light. The development of processing plants would be dependent on the
local price of imported fuel which is moderate. Pilot trials have been
completed on Tanna.

The Land. Evaluation of some geothermal sites has been initiated. These
are MW resources. Thus the proximity of large population centres provides
the most viable scenario, or the introduction of large industries.

EXPLOITATION

The resources are discussed for their potential to provide various levels
of dependable energy, examples being cited for the central and southern
islands.

Wind. Being the major reason for the visit this resource was addressed in
detail. Wind speeds indicated from the long term records at Port Vila are
not promising.   However records for the surrounding region (US Naval
Weather 1979), suggest that the regional mean value of 3.5m/s is too low.
Adjacent New Caledonia, Fiji and Wallis, (all lying within a 1500km radius
in open sea without significant intervening sea currents) exhibit equivalent mean winds of 5.0 and 5.4 and 5.7m/s respectively. There is a
significant probability of hurricane force winds occurring during the wet
season, entailing a design wind limit of 70m/s for machines.

Wind pumping is certainly viable and exposed sites are likely candidates
for electrical generation. With a water need identified and experience
with water pumps at several sites on Efate, their reintroduction on a wide
scale would be beneficial. Tests siting turbines remote from well heads
are shown to be successful, (Goezinne and Eilering, 1984, Fritzsche and
others, 1984) with the advantage of improved wind fetch and access to
saline free water, Fig.2.   These workers addressed electrical and
hydraulic connections to the pump. The former method would provide the
facility to charge batteries, energising community lighting and for
vaccine cool stores.

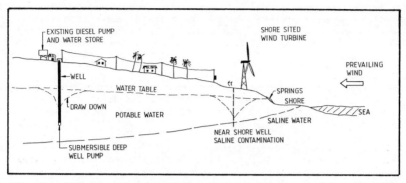

Fig. 2. Installation of separate wind generator and deep well pump

Garside and Peacock,(1981) assessed two turbines of 1.8 and 3.5kW capacity and cut in speeds of 4.0 and 5.0m/s respectively, and showed that the 3.5m/s wind regime would provide an annual output of 3 MWh with the capability of providing 0.5kW for 25% of the year.

Large farms of wind turbines developed in the USA, supply significant electrical energy to the local utilities. Some recent sites developed exploit modest wind speeds and machines of 300kW capacity (Wind Directions, 1985) have been developed for these sites. There is therefore proven hardware for use at the more exposed sites in Vanuatu able to operate into a utility network, such as Port Vila and Luganville. The former region has two possible sites, Pango Point, with existing power cables adjacent and Devil's Point, entailing additional lines. In both cases the energy generated would supplement that generated by the diesel stations.

The highest winds occur from 0900 until 1600 hours based on 10m. height observations, (there may be more continuous winds at increased heights). The large settlements use significant air conditioning, giving a useful match of resource and demand. To compensate for village demand/energy imbalance storage media would be necessary, which may exist as a water supply tank and could include batteries to power lighting and cool stores.

Fishing vessels were wind powered but many have been replaced by powered boats with the associated fuel and engine maintenance costs. For improved local input the provision of wind power even for larger vessels would improve the local and export market prospects. Retraining of the local fleet operators is underway in order to provide more domestic fish and sail boat operation training could be introduced concurrently. A sail assisted ship is operating successfully within the islands using a local crew.

Solar. Insolation in the latitudes of Vanuatu is greater than 800 W/m^2 and yearly average sunlight is 2100 hours at Port Vila and 1700 on Santo, November and December being the most sunny months.

Solar cells are used for battery charging at strategic radio transmitters throughout the group, but array costs are still prohibitive for general use. Passive solar panels for water heating are not common. Indigenous

units built using wooden framing have failed prematurely in the tropical
conditions. Aluminium extrusions and varieties of glass are available for
an entrepreneur to produce local units. In 1981 however imported LPG
provided the cheaper option for water heating in town situations. A few
medical establishments used Australian units. Local manufacture would
expand such usage.

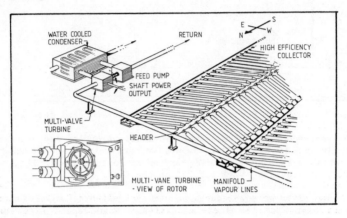

Fig. 3. Solar powered Rankine cycle irrigation pump

Cranfield is active harnessing the UK sun, to develop prototype solar
powered prime movers, cold stores and crop dryers. A Rankine cycle using
high efficiency solar collectors and a multivane expander provides shaft
power suitable for irrigation purposes (Fig. 3),(Hussein and others,
1981); shaft energy could power an electrical generator. Similar solar
collectors provide a cooling requirement using an Electrolux unit;
temperatures of $-12^{o}C$ have been achieved in a $32^{o}C$ environment. With
suitable insulation long term vaccine and drug storage looks feasible at
moderate cost, (Uppal and others, 1985). Copra, the main export, is dried
in significant quantities using wood driers. The potential of using solar
driers was not explored.

Wood. Dark bush is an abundant energy source, currently used by the
village population accounting for 35% of the Republic's energy
consumption. The resource is not always near the centres of population
but conversion of wood to charcoal undertaken at some centres would
provide a light concentrated fuel source for wider economic distribution.
A charcoal industry expanded in rural areas has the double advantage of a
localised fuel complemented by improved cash flow.

The potential exists for an energy forest where timber converted to
charcoal could provide the heat source for substantial power generation.
This would be a longer term venture, with the planting of fast growing
varieties for use in improved technology small generation systems
(Duckworth, 1985).

Some prototype running has been completed using charcoal generated
producer gas to power diesel engines in the teaching centre. Schools,
hospitals and tourist residences in remoter regions use diesel power for
their electrical needs. Producer gas systems will be most efficient for

continuous power demands.

Geothermal. Three potential sites were visited: Tanna, Efate and Tongoa.
Four other sites exist in the Northern islands. The sources exhibited
steam, hot surface water and large areas of hot ground and have the
potential to provide energy far in excess of the needs of the Republic.
None of the areas are expected to exhibit more violent activity, but...
The Efate source is away from the main town, but could be a future
industrial power source away from main population centres - aluminium
smelting has been considered.

The Rankine cycle system discussed earlier would operate using a
geothermal source. With the proximity of villages on Tanna and Tongoa,
there is potential for continuous power at small levels - heat sources are
near the sea to provide the cool sink.

Hydro. Large scale production of electricity has been investigated at
river sites adjacent to the main population centres. The probability of
earthquakes is finite and could rule out large installations. Foundations
present a problem in the friable bedrock.

Small scale installations (5 - 20kW) would be feasible at sites where
village and river were adjacent. Seven such sites exist. Hydro installations in common with wind are expensive to install but have long lives
and provide a constant power source, (Watt Committee, 1985).

Wave. The UK work has slowed recently, but the ideas generated have some
application in Vanuatu, sadly without British involvement. The Norwegians
have been studying nearby Fiji and suggest single oscillating water
columns installed facing the prevailing swell mounted at reef edges, Fig.
4 (Steen, 1985), providing a peak output of 500kW. An alternative tapered
channel arrangement focuses the swell into a reservoir for later
discharge through a low head turbine with a capacity of 350kW. This is a
variation on a scheme proposed for Mauritius, which uses a wall mounted on

Fig.4. Reef mounted oscillating water column generator (Steen, 1985)

the reef edge. The Norwegian schemes are compact, suitable for the larger
island towns having an established supply. Santo may benefit from such
units facing into the prevailing swell but some attenuation may occur
through the islands. Vila is sited on the lee of Efate, away from the
prevailing swell.

Ocean Thermal Energy Conversion. Experimental work is continuing and the
status of application in the Republic will need to be reviewed. As noted
earlier potential sites are adjacent.

Plant and Animal Waste. Because the traditional methods of grazing and
tending do not confine the animals to one place, village scale production
is not feasible and attempts have failed. For any success a 200 pig unit
is necessary, which includes few farms in the Republic and does not help
village communities.

Plant Oils and Fermentables. Schemes for these products have been
planned, but the costs remain high. A local oil mill in Efate
successfully uses low grade oil to power the diesel. Lower fuel oil prices
are not a spur to find alternatives but local currency exchange rates
colour the real price of fuel.

A NEGLECTED FACTOR

This review of the energy scenario and the means of harnessing alternative
sources from within the Republic is inevitably coloured by a European
view. A recent article in New Scientist (Charnock, 1985), addressed the
application of alternative technologies and discussed some unsuccessful
projects where local needs had not been identified. For example, cooking
methods have been subject to worldwide studies improving safety and
efficiency and easing the work load. Applications in Nepal generally
succeeded in these goals, but the reduced smoke allowed insects into
dwellings and reduced light necessitated introduction of a light source.

Vanuatu includes developed towns with European influences, where conventional solutions would be applicable. However, 80% of the population live
in a village situation where a drinking water supply is paramount,and
other specific requirements need to be identified. A fundamental section
therefore of any energy programme should establish true needs of town,
village, school or individual before entering the detailed evaluation
stage.

SOME CONCLUSIONS

Bearing in mind the scope of the study it is concluded that :

o   Water pumping using wind power is economic in many parts of the
         Republic.

o   Where wind speeds are in excess of 3.6m/s electrical generation
         is economic using equipment at £7000 (1981 prices), with fuel oil
         at the prevailing price.

o   Solar and geothermal resources could be economically exploited to
         provide shaft power using engines currently under development.

o   Hot fluids could be readily provided from passive solar units built using locally available materials.

o   Abundant wood reserves could provide the basis for a charcoal
        industry and demonstrates the potential for an energy forest.

o   Hydro energy resources needed further collation and assessment.

o   Recent wave energy proposals for adjacent countries hold promise
        for application in Vanuatu.

ACKNOWLEDGEMENT

This study was sponsored by the UK Overseas Development Ministry.   The
assistance of the Rev. K.C. Calvert was an essential part of the study.

REFERENCES

Calvert, K. (1980). The nature and function of the Kristian Institute
Technology of Weasisi, January 1980.

Charnock, A.(1985). Appropriate technology goes to the market, New
Scientist, May 1985.

CPO,(Central Planning Office),(1980/1). Government of Vanuatu, Report
1980/1.

Duckworth,P.A. (1985), Rural wood burning power stations, Seminar
I.Mech.E., London, Package power stations for export, 7th November 1985.

Garside, A.J., and Peacock, R.E.(1981). A report on the use of wind
energy and other alternative energy sources on the islands of Vanuatu, ODA
Report.

Goezzinne, F. and Eilering, F.(1984). Water pumping windmills with
electrical transmission, Wind Engineering, Vol.8, No.3, 1984.

Hussein, M., O'Callaghan, P.W. and Probert, S.D. (1981). Solar activated
power generator utilising a multi vane expander as a prime mover in an
organic Rankine cycle. Solar World Forum, ISES, Brighton, UK, 1981.

Uppal, A.H., Norton, B. and Probert, S.D. (1985). Low cost solar energy
simulated refrigerator for vaccine storage. The Engineer (accepted for
publication).

US Naval Weather Service Detachment,(1979). Summary of Synoptic
Meteorological Observations, September 1979. (AD-A073 446).

The Watt Committee on Energy (1985). Small scale hydro power. Report
No.15, March 1985.

Wind Directions,(1981). UK News, April.

Steen, J.E. (1985). Norwegian project : energy from the waves in the South
Pacific. Scandinavian Energy No.1, 1985.

System Analysis of a Solar- and Windpower Plant for Nordic Remote Locations

H. C. Rasmussen

Niels Bohrs Allé 25, DK-5230 Odense M, Denmark

ABSTRACT

Nordic remote locations have the possibility to be supplied nearly entirely
with energy from local resources. For a municipality (Hasle) on the island
of Bornholm (50,000 inhabitants) many initiatives have been taken to establish renewable energy plants combining the supply of heat, power and transport energy demands. A few of the demonstration plants are already in operation but most still lack appropriate financing means. Some of the plants
are described in this work; these are a solar (PV) - and windpower plant,
a collective biogas plant, a combined heat, power and transport energy biogas plant for an agricultural farm, and an electric vehicle for the town of
Hasle. Thus a Renewable Energy - Combined Heat, Power and Transport (RE-
CHPT) concept is under development. Economic analyses are given for the
solar- and wind power plant and for the collective biogas plant.

KEYWORDS
Photovoltaics, wind, biogas, heat, electric power, transport energy.

BACKGROUND

On the island of Bornholm (50,000 inhabitants) the present energy supply is
based on petrol, fuel oil and electricity form the Danish mainland. Electricity production from renewable energy sources is foreseen to be economically
attractive earlier in the technico-economic development process than in the
Danish mainland due to

- special emphasis is put to supply security thus a standby power station is kept operational for emergency situations
- the electricity demand on Bornholm is much more evenly distributed over the
  year than in most of Denmark and the daily variations are slightly larger
  than in Denmark as a whole due partly to a weak industry and partly to
  the many tourists visiting the island in the summer time.
  The former being shown in fig. 1.

A8.1

45

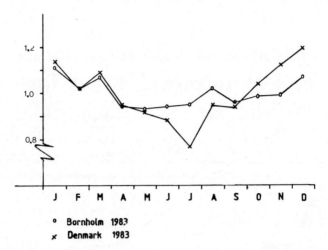

° Bornholm 1983
x Denmark 1983

Fig. 1. Monthly variation in the electricity demand on Bornholm and in Denmark (1983). The total demands being 642 TJ/yr, 87.5 PJ/yr respectively.
Sources: DEF (1984), ØSTKRAFT (1984).

The County council has initia ted the project "Renewable Energy on Bornholm" to achieve a more rational use of energy in the county and to develop
the local energy resources (straw, woodchips, urban waste, agricultural
waste, solar- and wind energy) thus reducing the dependence on imported
fuels. This project is further described by Jørgensen (1985).

ELECTRICITY SUPPLY ON BORNHOLM

Electricity is distributed on the island by "ØSTKRAFT" – a cooperation of
the municipalities. Until 1980 all electricity production took place within
the island, resulting in consumer prices of up to 180% of the price in the
main parts of Denmark. A cable has now been established, so that the price
difference is very much reduced.For the small island the production units
are smaller and more expensive to run than the big coal fired power stations
which are used on the Danish mainland. Some of the production takes place
with fueloil dieselmachines. Since local production is more expensive than
imported electricty, only 4% of the yearly supply in 1983 of 642 TJ was
produced locally. The peak demand is slightly increasing being around 35 MW
(1982).The cable down time was (1982) scheduled 9h 33 min. and unscheduled 18
min.

The import price is very dependent on the time of the day and on the supply
situation in whole Scandinavia, in 1982 ranging from 5.154 øre/kWh to 46.905
øre/kWh with an average of 22.1 øre/kWh.

The consumer price was for the first 100 MWh/yr 55 øre/kWh and for the exceeding consumption 48 øre/kWh excl of Danish energy tax of 15.5 øre/kWh
and excl of VAT of 22%. The consumers are also charged fixed 4560 Dkr/yr.

SITING OF HASLE RE-PLANTS.

Apart from the RE-plants being further described in the following until now
a windpark (10 x "100kW") and a biomass supplied district heating system
has
been designed in the municipality. The proposed sitings being shown in fig. 2.

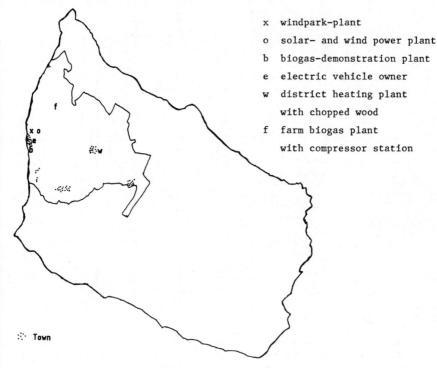

x   windpark-plant

o   solar- and wind power plant

b   biogas-demonstration plant

e   electric vehicle owner

w   district heating plant
    with chopped wood

f   farm biogas plant
    with compressor station

Fig.2. RE plants in the municipality of Hasle.

SOLAR- AND WINDPOWER PLANT

A calculation model of solar- and windpower plant has been set up to investigate the extent to which a photovoltaic plant together with wind energy
better can match the electricity load curve. For the revealed best combination of photovoltaic- and wind electricity production-units the yearly
electricity production under normal conditions has been calculated.

The best system is composed of
- a photovoltaic production unit built with solar cell arrays in one field
  with a nominal power production of 36 kWp (AM 1.5)
- a windpower production unit equipped with 200 kW generator capacity
- a battery storage system of 200 kWh
- a DC/AC inverter, power conditioning system and a super control system

A diagram of the combined plant is shown in fig. 3.

48   H. C. Rasmussen

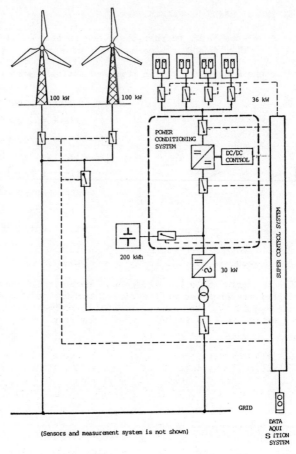

Fig. 3. Solar- and windpower plant. Block diagram   showing components and
energy and information links.

Annual Energy Balance for the plant (norm year)
- origin wind energy   1656 GJ/y   (460 MWh/y)
- origin solar energy   130 –   (36 MWh/y)
Total   1786 GJ/y   (496 MWh/y)

Economic Analysis

To urge local technological development and improve local economy and
employment 1/4 of the PV-panels are foreseen to be produced in Denmark,
with Bornholm subcontractors, also the two windmills will be produced on
Bornholm.

Lifetime   Windmills and PV-plant   20 y
                            Batteries   10 y
                            Electronics   20 y

A8.4

Investment costs

Total investment excluding preliminary investigations and planning, measuring system and measurements (excl. VAT) 7,900,000 Dkr (1985) with a net interest rate of 5.5% the annual payments are 671,000 Dkr/yr (incl. replacement of batteries after 10 yrs). Per unit of energy: 1.35 Dkr/kWh.

Operating costs (per y)

Excess cost at the Municipality of Hasle, Technical Department	20,000 Dkr
Skilled and specialist assistance	30,000 -
Insurance (0.6 - 1.0 % dependent on component)	50,000 -
Total	100,000 Dkr

The total electricity prices then will be   1.55 Dkr/kWh

Scenarios

Taking an increase in the electricity prices of 1.5% p.a. in real terms the average price over 20 yrs is 0.75 Dkr/kWh leading to a simple payback time of 25 yrs.
The plant will by including a battery bank increase the supply security on Bornholm. Electricity produced at the diesel machine at "ØSTKRAFT" will be at approximately 1.50 Dkr/kWh, (excl. capital investment) but there is not any agreed objective mean to include this in the economic evaluation of the RE-plant.

The future outlook for this type of plant for remote locations at the year 2000 can optimistically be described by capital investment 5.9 M.Dkr, O&M = 100,000 Dkr/yr, the cost of electricity produced y diesel machines inclusive of capital costs and O&M 2.60 Dkr/kWh giving a simple payback time of 5 yrs.

COLLECTIVE BIOGAS PLANT

A thermophilic biogas plant is to be situated just outside the town of Hasle, and large residue tanks - 3-4000 m^3 - for de-gassed biomass are placed appropriately in proportion to the surrounding farms. At the farms small mixing tanks shall be established - if not already present - for collecting the liquid and dry organic waste. This organic waste is transported from the individual farms to the receiver tank of the biogas plant by means of a 20 tons tanker driven by a freight-trucker, operated by the personnel of the biogas plant. The same system brings the de-gassed organic waste from the residue tank of the biogas plant to large decentralized residue storage tanks.

At the biogas plant a high pressure compressor with a purification system is situated for cleaning the biogas and storing it in a storage tank battery at 300 bar. The storage tank battery is transportable and is foreseen transported daily from the biogas plant to the existing depot at a nearby town where a filling station is installed, and CBG (compressed biogas) buses (storage tank 165 bar, 700 l) and vans are fuelled and serviced, substituting 100 tonnes of diesel oil/y . Furthermore, a 700 m long pipeline leads the biogas from the plant to the town of Hasle to be used for heating purposes. Substituting 474 tonnes of fuel oil/y .The total combined system is shown in fig. 4.

A8.5

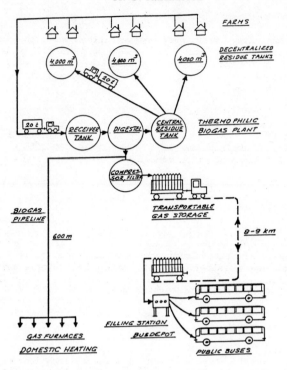

Fig. 4. Central biogas plant for heating and transport energy supply

The gas production is estimated to be 1824 m^3/day where 10% is used for the biogas process giving a net production of 1640m^3/day. The plant will operate 350 days/y giving 574000 Nm^3 methane/y .

Operation Economics

The obtainable fuel price will be 100% of the oilprice (3200 Dkr/ton) for public customers and 95% of the oilprice for private customers.

This leads to the following operation economics:

Income from gas	1,772,000 Dkr/y
Increased fertilizer value	400,000 Dkr/y
Total income	2,172,000 Dkr/y
Annual operation and maintenance cost	: 600,000 Dkr/y
Interest and repayments	: 625,000 Dkr/y
Production costs	1,225,000 Dkr/y

INDIVIDUAL RE-CHPT SYSTEMS

For an agricultural farm a biogas plant has been designed to supply the farm with not only electricity and heat but also with energy for one of the tractors operating in the fields.

The system components are:

1. A biogas plant of 250 Nm^3 biogas/day, 60% methane
2. A biogas engine (dual fuel with turbo charger), 20 kW_e, 40 kW_{th} total efficiency 0.83
3. A biogas boiler
4. A biogas cleaning and compressorstation (200 bar) and storage of 3 x 50 l.
5. A tractor (100 HP) with a gas-dual fuel engine.

TRANSPORT

To demonstrate further that on the island electric vehicles can cover an essential part of the transport work due to the limited geographical distances an ordinary Fiat 127 Fiorino van has been rebuilt for the Municipality of Hasle. The van is shown in fig.5. The normal drive train (with Otto-engine) has been exchanged with an electric drivetrain consisting of an electric traction motor, chopper and battery.

Fig. 5. The Fiat 127 Fiorino electric van rebuilt for the municipality of Hasle by the Energy Research Laboratory, Odense.

CONCLUSION

On the island of Bornholm the inhabitants as well as visitors in a few years time might well see a novel energy system appear where the main features are energy savings and energy production from local resources in combined heat, power and transport modes (CHPT-systems)

A8.7

REFERENCES

DEF, 1984: "Dansk Elforsyning-Statistik 1983" Danske Elværkers Forening,
  Copenhagen
ØSTKRAFT 1984: "Beretning 1983". Rønne Bornholm
K. Jørgensen, 1985: "Renewable Energy at Bornholm - self sufficiency
  for an island community in Denmark" to be presented at "Energy for Rural
  and Island communities (ERIC 4) Inverness Sep 16-20, 1985
Energisekretariatet, 1985: "Oplæg til VE-udbygningsplan 1985-90. Projekt
  for vedvarende Energi på Bornholm." Bornholms amtskommune.

ACKNOWLEDGEMENT

The information for this work has been provided by the Energy Research
Laboratory and the EEC demonstration project in the field of Energy Savings
SE-555-84.

TOPIC B
Photovoltaic Electricity Generation

Appraisal of Photovoltaic Power Systems for Rural Development in Nigeria

A. O. Odukwe and K. Mukhopadhyay

National Centre for Energy Research and Development,
University of Nigeria, Nsukka, Nigeria

ABSTRACT

The energy needs in Nigeria are rising in all sectors whereas the supply seems
to be still heavily dependent on oil and electricity generation from NEPA
(the National Electric Power Authority).   Studies reveal that electricity
generation is far from adequate especially in rural areas where distribution
presents a heavy burden on the depleting economy.

In the light of these problems and socio-economic situation of the Nigerian
population, solar energy is considered as a worthwhile augmental energy source.
In considering the demand and supply system, economy, social and cultural
set up of the rural communities where over 80% of the Nigerian people live,
a case study was made of Iheakpu-Awka. The result showed that a photovoltaic
power system is quite attractive. On further consideration of three approaches
(viz integrated centralized, discrete centralized and roof-top) the integrated
centralised approach is recommended.

KEYWORDS

Decentralised energy;   integrated centralised energy;   photovoltaics.

INTRODUCTION

Nigeria, though blessed with natural resources like oil, bright sunshine all
year, coal, gas, hydropower, forestry etc. has an economy that is almost 90%
dependent on oil; thus making her economy prone to fluctuations in the
world oil market. Moreover, the neglect of agricultural and non-urban
development has led to the emergence of an abnormal social structure.   With
80% of her population living in vast non-urban areas, a large part of Nigeria's
gross domestic product can emanate from the rural sector if appropriate
measures are taken.   However, this will be a function of the available energy.

According to analysis made by (NEPA), the country experienced a total shortfall in energy of 4,217 GWh in 1984, with the vast rural sector almost neglected.
Over 45% of the total energy sales of the country are consumed in Lagos alone.
Projected analysis by NEPA predicts a shortfall of 7050 GWh in 1990.   This
includes the rural sector. Consequently, efforts should be made to create

55

B1.1

new power plants to meet this shortfall. For a real break-through in agricultural and non-urban development, small scale power plants with minimal
operational and maintenance costs are imperative. Also for obvious economic
reasons, a locally available energy source is important. This makes solar
energy as an alternative very attractive as every part of Nigeria enjoys
bright sunshine, about 2300 kWh/m^2 on a horizontal surface (2), throughout
the year. Thus harnessing solar energy to meet the electrical, mechanical
and thermal energy demands of the rural sector would help boost agricultural
production and improve the socio-economic structure.

PHOTOVOLTAIC POWER SYSTEMS (P.P.S)

Figure 1 shows the schematic diagram of a stand alone photovoltaic power
system. The solar array consists of series and parallel connection of a
number of photovoltaic modules which produce direct current when solar insolation is available. One square metre of a fixed solar photovoltaic array
will yield 500-550 Watts-hrs of electrical energy in a 24 hour daily cycle.
The anticipated life of the modules is 20-30 years, during which very little
maintenance like cleaning of the surface, external connection between modules
etc. are to be considered. The output from the modules is proportional to
incident illumination. Thus there is a need for battery storage of electricity in order to supply power in case of low or zero illumination intensity
(rainstorm or night). A steady supply of power to the load is thus ensured.

Power conditioning equipment may be required for:
i)   protecting storage batteries.
ii)   converting DC into AC if the system is to be integrated with AC equipmen
iii)   extracting the maximum power that solar cells can generate.

The power conditioning equipment consists of voltage regulators, blocking
diodes, DC to DC transformers, DC to AC inverters, maximum power point
tracker, etc.

The load is dependent on the nature of applications. For rural applications,
it may consist of domestic and street lighting, household appliances, water
supply systems, small scale industries, etc. All these components of PPS
are to be evaluated for a particular location.

The PPS could be designed in any of the following ways:

(i)   Integrated centralised approach, in which the same solar array
       generates power for the whole community in a manner similar to a
       central power station.

(ii)   Discrete centralised approach, in which separate solar arrays are
       used for the different components of the community 's electrical
       load. For instance, an array can be used solely for lighting while
       another is used for appliances, water supply etc.

(iii) Decentralised or roof-top approach, whereby each household is provided
       with its own solar panel in a manner similar to a small electric
       power generating set.

A CASE STUDY FOR PPS FOR A RURAL COMMUNITY IN NIGERIA

In order to design PPS for typical rural communities, the electrical energy
requirements of a number of such communities in different parts of Anambra
State, Nigeria were surveyed.

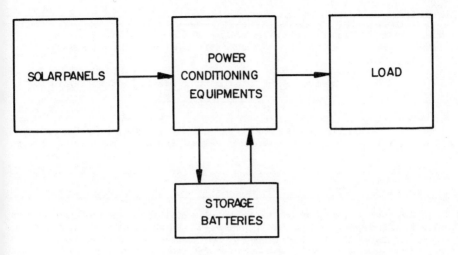

Fig. 1   Schematic diagram of stand alone solar
         photovoltaic electric power systems

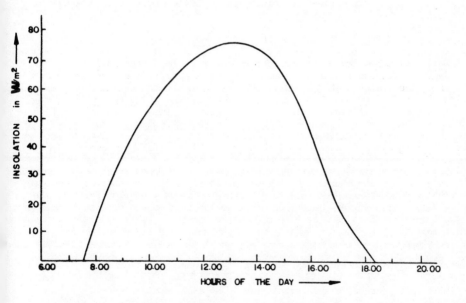

Fig. 2   Average daily insolation at Nsukka, Nigeria

General Description of the Location

After a thorough survey, Iheakpu-Awka in Igboeze Local Government Area of
Anambra State, Nigeria; was chosen as a study case.   Table 1 gives the
statistics of available facilities and estimated land area, population etc.
of this community.

TABLE 1   Communal Utilities and Structure of Iheakpu-Awka
Rural Community

Area Population (km^2)		No.of Villages	No.of Markets	No. of Health Centres	No. of Community Halls	No.of Primary Schools	No.of Post- Primary Schools	Av.No. of Persons per House- hold	Av.No. of Houses
26	12,538	12	4	1	1	6	1	10	1250

Table 2 gives a summary of the survey results of this community based on data
collection through questionnaires, interviews and consultations with villagers
and officials of the local government.   No reliable data were available for
cooking because almost all the houses use firewood.

From the table, water consumption is very low because of the absence of any
natural source and non-connection of the community to urban water works.   The
villagers either buy water from tanker owners at $2.50 per 40 gallons or trek
10 km to the nearest water source.   The depth of the water table in this area
is about 200m (1).   Although positive displacement pumps, such as the jack
pump, are capable of pumping water at a low discharge rate with lifts up to
300 m   , it is not very promising to use such pumps for a community of
12,000.

TABLE 2   Summary of the Survey Results

Daily per Capita water con- sumption (litre)	Daily per Household kerosene consumption (litre)	Percentage of Households that own T.V. Refrig- Radio/ erators Cassette Players			Estimated daily electrical energy need for lighting appliances (kWh) Per Per Household Capita		Water Level (m)
8	0.3	4	4	58	6.2	0.64	180-200

Food crops are mostly cultivated with a very high dependence on the seasonal
rains (from March/April to October) for irrigation.   With a very low water
table (180-200m), unreliable farming data and a very poor estimate of the
widely scattered farm lands, farm irrigation using photovoltaics is not very
attractive and so has not been considered here.   Finally, the electric energy
requirements of the community's civic centres, school and health centre have
been considered along with those for the households.

Analysis of Electrical Energy Needs and Load Patterns

With 1250 households, electricity would be required for:

(a) domestic and community lighting
(b) household appliances such as television sets, refrigerators and electric
    irons, which are currently being powered by car batteries or diesel genera

About 60% of the households are willing to afford the materials and overhead requirements for electrification. For domestic lighting six 40W lamps per household are required. Six community T.V. centres are proposed which will contain six T.V. sets and twelve 40W lamps. The health centre requires two refrigerators and ten 40W lamps. At present solar powered refrigerators are available in the market (9,10).   Over a 24 hour period, the refrigerator motor will operate 30% of the time. Finally a total of 5km of road will require about 100 lighting posts.

Table 3(a) gives the analysis of the electricity requirement of the community using the integrated centralised approach.

Table 3(b) gives the analysis for discrete centralised while Table 3(c) gives that for the decentralised roof-top approach.

TABLE 3   Electrical Energy Analysis and Load Pattern for
Iheakpu-Awka Rural Communities

(a)   Integrated Centralised Approach

Details of Category	Average Daily use (hour)	Power (kW)	Energy (kWh)
Day Time Load			
(i) 475 Radio/Cassette Player sets (25 W/set)	One	11.875	11.88
(ii) 34 Refrigerators (127 litre) (60W)	Four	4.08	8.16
(iii) 34 Electrical Iron s (750 W/unit)	Half	25.50	12.75
Total Day Time Load (E_D) =			32.79
Night Time Load			
(i) Domestic lighting (six, 40W lamps/household)	Three	195.60	586.80
ii) 34 Refrigerators	Three and a half	2.04	7.14
iii) 40 T.V. Sets (100W/set)	Four	4.00	16.00
iv) 475 Radio Cassette Players	Three	11.88	35.64
v) Street lighting	Four	4.00	16.00
Total Night Time Load =			661.58
Total Load $(E_D + E_N)$ =			694.37

(b)   Discrete Decentralised Approach

Unit	Details of the Category	Time Segments	Power (kW)	Energy (kWh)
I	Day Time Load (E_D)	0	0	0
Lighting	Night Time Load (E_N)			
	(i) Domestic Lighting	Three	195.60	586.80
	(ii) Street lighting	Four	4.00	16.00
	Total Night Time Load =			604.80
	Total Load $(E_D + E_N)$ =			604.80

B1.4

(Discrete Decentralised Approach cont'd

II   Day Time Load
Appliances (i)   475 Radio/Cassette Player
                          Sets   One   11.88   11.88
            (ii)   34 Refrigerators   Four   2.04   8.16
            (iii)   34 Electric Irons   Half   25.50   12.75

Total Day Time Load (E_D)   =   32.99

            Night Time Load
            (i)   40 T.V. sets   Four   4.00   16.00
            (ii)   475 Radio/Cassette Player sets   Three   11.88   35.64
            (iii)   34 Refrigerators   Three/half   2.04   7.14

Total Night Time Load (E_N) =   58.78
Total Load ($E_D + E_N$)   =   91.57

            (c)   Decentralised or Roof-top approach

Details of Category	Time Segments	Power (kW)	Energy (kWh)
Day Time Load			
(i) 1 Radio/Cassette Player set	One	0.025	0.025
(ii) 1 Refrigerator	Four	0.06	0.24
(iii) 1 Electric Iron	Half	0.75	0.375
Total Day Time Load (E_D) =			0.665
Night Time Load			
(i) 6x40W lamps	Three	0.24	0.72
(ii) 1 Refrigerator	Three/half	0.06	0.21
(iii) 1 T.V. Set	Four	0.10	0.40
(iv) 1 Radio/Cassette Player set	Three	0.025	0.075
Total Night Time Load (E_N)			1.405
Total Load per Household ($E_D + E_N$) =			2.07

Methodology and Design of PPS

Precise design requires accurate insolation data, subsystems configuration
and their efficiencies.   Detailed design procedures (6) and less precise
but reasonably accurate and simplified design methods (10) are available.
Nsukka insolation data (Fig.2) (4,5) were used as an approximation to that of
the chosen location   by virtue of their proximity. A maximum of three days'
storage can be considered. Finally, the sizes of the components can be
computed using the energy balance condition (10).   Using the balance of
system (BOS) costs consideration, and from the nature of communities under
study, the integrated centralised approach is suitable for Nigerian
applications. Also, based on the work done in Upper-Volta, West Africa, by
NASA-LEWIS Research Centre, the size of the BOS/peak Watt (R in U.S.$)
decreased with size (S) of the installation as (3)

$$R = 296 \ (S)^{-0.412}$$

B1.5

Therefore, the PPS will be cheapest and better maintained in the integrated centralised approach which employs the biggest capacity.

CONCLUSION

Energy and the Nigerian economy are almost totally dependent on fossil fuel. Uses of conventional sources of energy except coal are on the increase due to an uncoordinated energy plan. The attendant problems of over-dependence on oil can be solved if the country's bright sunshine, huge natural gas and forest reserves are properly harnessed. With a perennial shortfall in electricity generation, the virtually non-existing supply of power to the rural sector, which is vital to Nigeria's socio-economic survival, will continue. Consequently, the possible application of Photovoltaic Power Systems for this purpose has been presented. Of the three approaches evaluated, the integrated centralised approach is most suitable for Nigeria. Finally, the present state of art of research and development, warrants technical and socio-economic studies of actual installations.

REFERENCES

1. Adegoke, O.S. (ed), (1980).   Shagamu Quarry and Bituminous Sands of Ondo and Ogun States. Nigerian Mining and Geosciences Society Special.
2, Berkoviteh, I. (1981). Trapping Nigeria's Solar Energy. West African Technical Review, 31.
3. Dickinson, W.C. and P.N. Chermisinoff (eds), (1980). Solar Energy Technology Handbook, Part B: Appl.,System Design and Economics, Marcel Decker Inc., New York. p.328.
4. Ezekwe, C.I. and C.C.O. Ezeilo. Measured Solar Radiation in a Nigerian Environment compared with Predicted Data. Solar Energy, 26, 181-186.
5. Ezeoke, A.S.O., (1979). Solar Insolation in Nsukka, B.Eng.Proj.Rep.UNN.
6. Imamura, M.S. et al., (1976). Final Report MCR-76-394 under contract NAS 3-19768, Martin Marietta Corp.
7. National Electric Power Authority (1982). NEPA in the 80's.
8. N.E.P.A. (1982). Power System Dev. Plan, 1982-1990.
9. Nolfi, J.R., (1981). Alt.Energy Appl. in Nigeria. West African Tech. Rev.
10. Saha, H., (1981). Design of Photovoltaic Electric Power System for an Indian village. Solar Energy,Vol.27, No.2, 103-107.
11, Watkins, T.L., Solarex Corporation (Private Communications).

Plans for a Demonstration Photovoltaic/Wind/Pumped Storage System Integrated on an Island in Boston Harbor

M. Westgate* and D. A. Bergman**

*Education and Resources Group, Old City Hall, 45 School Street, Boston, MA 02108 (617) 523–8678, USA
**Northeast Regional Experiment Station, Massachusetts Institute of Technology Energy Laboratory, Concord, MA, USA

ABSTRACT

This paper describes the energy section of the Master Plan for an Island in Boston Harbor devoted to public educational and recreational use. The Island provides an unusual opportunity to demonstrate the viability of various alternative energy systems, integrated to provide power for 177,000 square feet of existing buildings, built of brick in 1904, abandoned in 1947, with rehabilitation begun in 1983.

Placing initial and primary emphasis on photovoltaic cells, the plan also utilizes windpower and water pumped storage. Diesel generators are to be installed as backup.

Educational displays are to be interactive with the public, demonstrating power availability and use and the value of conservation.

Island visitation is projected at 20,000 initially, growing to 200,000 annually. Electrical demand (currently there is none) would grow from 28,000 KWH to 250,000 KWH annually.

Peddocks Island offers the advantages of a remote island location combined with easy access (50 minutes by boat) from downtown Boston, its universities, corporations and international airport.

The Island is a potential showpiece for the Boston area's 2,800,000 residents and for visitors from the United States and overseas.

KEYWORDS

Photovoltaics; wind; water pumped storage; integrated system; demonstration project; island; Boston Harbor.

B2.1

The overall goals and objectives for Peddocks Island[1]in Boston
Harbor are:

°To open the Island resource to public recreation and education
 with special emphasis on marine resources, making it accessible by public transportation.

°To plan the Island's use in an ecologically sound manner with
 thoughtful utilization of natural resources consistent with
 conservation and protection.

°To facilitate the appropriate preservation of a number of the
 historic buildings on the Island.

°To provide opportunities for year-round use of the Island, and
 to promote use of the Island by people of all ages, sex, race
 and economic means, particularly from communities surrounding
 Boston Harbor.

°To encourage the development of income-generating activities
 to help propel the Trust toward self-sufficiency.

°To evaluate the effectiveness of the programs offered on a regular basis.

°To support exploratory, innovative programming including:

 * teaching and demonstrating the principles of economic and
   ecological self-reliance utilizing sustainable design (examples include energy-efficient and cost-conscious utility systems and day to day Island operations which emphasize reducing power and water consumption and protection of Peddocks'
   resources);

 * obtaining active participation by people of all ages and
   backgrounds in the programs, maintenance and revitalization
   of the Island, including development of a process for involvement in its management and protection;

 * using the Island as a laboratory for educational innovation
   fostering the interaction of teachers, administrators, community members and parents toward the improvement of instruc-
   tion (examples include expansion of the Hull High School model, weekend workshops, summer courses co-sponsored by Tufts
   University and other institutions, meeting areas offering an
   island retreat for interaction and problem solving, and opportunities for parents and children to learn together through
   activities like Community Boating's proposed sailing camps);

 * emphasizing hands-on learning (as part of the basic instruc-
   tional program of the schools) and student responsibility for
   the initiation and implementation of projects contributing to
   revitalization of the Island through a mechanism bringing
   adult and student, higher education and public schools, institutional and individual, corporate and volunteer resources
   together in shared tasks (examples include renovation of the
   buildings under the guidance of a resident community of re-
   tired master carpenters, cooperation with scientists and other university-level professionals to conduct research and im-

[1] as stated in Peddocks Island Master Plan prepared by Sasaki
    Associates and Peddocks Island Trust, January 1984

plement projects in such areas as erosion control, culturing
of marine animals, protection of archeological sites, and
public service projects such as the spring 1983 renovation of
the Island church by the Hancock Youth Group which demonstrates to the observers and participants the skills and concerns of young people).

°To provide opportunities for organized groups, families and
individuals to participate in Island oriented recreation, including boating, nature walks and other wholesome activities.

°To encourage organizations subleasing from the Trust to support the educational and recreational goals of the Trust, pro-
grammatically and financially.

°To maintain as many as possible of the current attributes of
an abandoned old military town, consistent with bringing new
life to the Island.

-----FERRY SERVICE   ⌐____⌐   1 km

ELECTRICITY

Since it is an educational goal of the Peddocks Island Trust to
promote self-reliance, several alternative schemes for generating electricity on the Island were examined. For cost comparison, the option of extending a subaqueous power cable from
Hull also was investigated. Regardless of the method initially
employed to obtain electricity, it is recommended that future
phases of Island development build on the experience gained during the first phase and seek to diversify energy sources.

Criteria For Electrical System

1. Practical, economical operating cost, transferable to household and small business contexts, modular, and easily expandable.
2. Educationally valid for development and application of hands
   -on skills for young people.
3. Capable of fostering an awareness of individual energy consumption levels, and promotion of self-reliance even in urban areas.

Anticipated Demand

The anticipated electrical demand by the end of the first phase
which includes one hostel at 16,500 square feet and a Visitors'
Center at 5,800 square feet, is 28,100 KWH/year. The anticipated demand upon completion of full development of the Island
is 250,000 KWH/year. These estimates are not conservative;
rather, they are based on the assumption that the management of
the Island will promote energy conservation.

Alternative Sources

Photovoltaic cells. Photovoltaic cells with a battery storage
and diesel generator backup could produce all the electricity
needed on the Island. Arrays of the cells could be placed
either on the roofs of individual buildings or at one, central
ground location possibly inside the sheep grazing fence. The
major constraint to using photovoltaics is the cost (currently
$10/peak watt, installed). However, recent reports indicate
that the cost may drop to $2/peak watt in the near future. Al-
so, the use of photovoltaics as the primary source of electricity on the Island will be more in keeping with the goals of the
Trust (i.e., self-reliance and education) than any option which
requires importation of electricity (i.e., extension of electrical service from Hull via subaqueous cable). If photovoltaic cells are adopted as the primary electricity source for the
Island, a generator should be purchased and installed for backup during extended cloudy periods.

Wind power. Small, wind power generators may be installed to
supplement other electricity sources on the Island. They are
not, however, recommended for use as the primary or sole source
of electricity. The life expectancies of wind power generators
are relatively short, resulting in a high annual cost per KWH
of electricity produced.

Pumped-storage hydro. Pumped-storage hydro is an option which
will allow storage of excess electrical power produced by other
sources. It is not, in and of itself, a true power source.
With such a system, excess power produced by solar and/or wind
power generators can be used to pump salt water up to a reservoir at the top of East Head. When supplemental power is required due to dark or windless conditions, this stored water
can be released to operate a low-head turbine. Due to the high
costs involved with such an installation, it is recommended

that this method of power storage be limited to a small (5 KW) educational demonstration and not relied on to replace battery storage.

Diesel generators. Diesel generators can be used to provide power for all anticipated Island needs. However, noise impacts associated with these generators render them undesirable as the primary power source. Therefore, it is recommended that diesel generators be installed only for emergency backup of other systems (i.e., wind or solar) and possibly for use at the Marine Science Building. Preliminary studies indicate the bunker area as a possible site for locating the generators.

Subaqueous cable to Hull. The Hull Municipal Light Plant must supply power to Peddocks Island if requested. A tie-in to this source will require installation of a subaqueous cable. This power source may be the most economical long-term solution to the power needs of the Island; however, it will not be in keeping with the goal of self-reliance.

COSTS

The anticipated costs (installation and operation) of each power source option, broken down to first year costs and ultimate costs when the Island is fully developed, are presented in the following table.

Power Option	First Year Cost Capital/Equivalent Annual[6]		Ultimate Cost Equivalent Annual[6]
Photovoltaic Cells[2]	$210,000	$32,000 ($1.15/KWH)	$ 58,000 ($0.39/KWH)
Wind Power[3]	$ 78,700	$22,000 ($0.78/KWH)	$117,000 ($0.78/KWH)
Diesel Generator[4]	$300,000	$49,500 ($1.76/KWH)	$190,000 ($1.27/KWH)
Subaqueous Cable[5]	$153,000	$19,500 ($0.69/KWH)	$ 45,000 ($0.30/KWH)
Pump-Storage (5 KW System)	$100,000	–	–

---

[2] Assumes: $10/watt of capacity in-place for first year, then a drop in price to $2/watt thereafter; maintenance at 1% of initial capital cost/year; battery/generator backup; 25-year life for PV cells; 4-year life for batteries; and 25-year life for generators. (The $2/watt cost is optimistic and dependent upon claims that advancing technology will decrease radically the cost of PV cells).

[3] Assumes: Installation cost of $3 per watt capacity, five-year life and battery and generator backup.

[4] Assumes use of three 300 KW generators.   Estimate includes
fuel cost.

[5] Initial capital cost of $150,000.   Assumes power costs inflate
at 5% above baseline inflation.

[6] All options are compared on a 25-year basis with an interest
rate of 12%.

Current energy costs for each power option are as follows:
photovoltaic cells 0; wind power 0; diesel generator $.05/KWH
for fuel only; subaqueous cable $.10/KWH for generated power
only (subject to substantial escalation).

Of the two most effective power options, photovoltaic cells and
subaqueous cable, the PV system has the advantage of permitting
phased development and therefore a gradual investment of capital.   A photovoltaic system is also likely, because of its innovative solution to Peddocks' energy needs, to attract foundation or public funding for demonstration projects.

INITIAL PHOTOVOLTAIC INSTALLATION

System Overview

The initial photovoltaic system for Peddocks Island was designed to meet the Island caretaker's electrical load as well
as provide fluorescent lighting for the Island church which was
to be used as a meeting hall.   The system was to be stand-alone
and modular in nature to facilitate further Island development.

The photovoltaic system for Peddocks Island was distinguished
from other small stand-alone systems in that it was to be the
beginning of a small Island grid.   We envisioned 120 Vdc power
lines disseminating from the battery bank housed in the church
shed.   The church was the logical choice for the array given
that it had a central location, large roof area (approximately
250 square meters), a relatively unobstructed field-of-view, an
acceptable orientation (tilt angle of 45° and azimuth angle of
−55°), and high visibility.   Dispersed small systems around the
Island were not given much consideration given that, with the
exception of the church, most of the buildings had slate roofs
(and roof-mounted systems were desired).   In addition, given
the amount of tree cover and the underlying philosophy to cut
as few trees as possible, dispersed small systems were not ten-
able.

Photovoltaic and Battery System

The initial PV system was to consist of 20 Mobil Ra25 modules
wired in 2 parallel strings of 10 modules in series to yield
120 Vdc.   The battery bank was to be composed of 10 Surrette
12 V / 140 A-hr batteries connected in series.   The battery
bank was designed to provide 16.8 KWH of storage at 120 V. (We
decided to design a system at 120 Vdc rather than the more conventional smaller voltages in order to minimize the wiring losses in the long cable runs from the church to the outlying
loads [from 250-350 meters].)   In addition, the system con-

tained a conventional battery regulator that performed voltage regulation.

Electrical Load

The church electrical load was to consist of four fluorescent lights used 2-3 days a month (non-winter months only) for approximately 2-3 hours per day. It was estimated that when in use the load would be approximately 200-300 Wh/day. The caretaker's electrical load would have consisted of a portable 12 Vdc radio, lighting in four rooms and a bathroom fan. This load was estimated to be approximately 1000 Wh/day.

System Attributes

The system had a number of attributes which were governed by system design constraints. For example, the proposed battery system, which was 120 Vdc permitted easy transfer to 120 Vac when the Island population had expanded sufficiently. Given that future construction on the Island was to be powered from the PV system, a 120 Vdc system permitted a number of power tools with universal motors to be run, whereas the more common 12 V system would have required the purchase of special tools. In addition, the array and batteries were expandable in integer multiples of ten which would allow for various load scenarios to develop on the Island.

Predicted System Energy Production

In order to predict the energy performance of the proposed system we utilized Boston Solmet Typical Meteorological Data and array temperature data from the Northeast Regional Experiment Station in Concord, Massachusetts. A simple model for performance corrections associated with array temperature was used and the daily energy values were adjusted for wiring and module mismatch losses as well as for battery inefficiencies. (It was assumed that two-thirds of the produced energy would be transferred through the battery bank which was taken to have approximately an 88% energy conversion efficiency.) In addition, the daily energy values were further reduced by 15% due to the fact that the array voltage would be set by the battery bank, thereby prohibiting maximum power tracking.

A plot of the system energy production and estimated consumption (assuming the entire load was downconverted to 12 Vdc) is displayed in Figure 1. The plot indicates that there are three months for which the PV system would not meet the load requirements. During these three months, the load as described above for the caretaker's house would need to be adjusted so as to accommodate the lower energy production. During the summer months there is an excess of energy produced on average. This would accommodate the additional seasonal load anticipated at the church. In addition it would have served the needs of a mobile workstation which could be serviced by either a long extension cord or a mobile battery pack charged by the system battery bank at night.

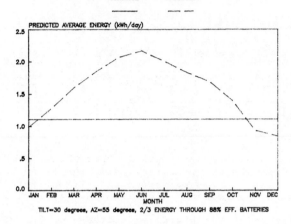

Fig. 1. System energy production and consumption.

WINDPOWER

A series of probably three 10KW windmills are to be located on
60 foot towers on top of the Island's 120 foot drumlin. Their
design and output projections will have the benefit of the operating experience provided on "Windmill Point" in Hull, ½ mile
away, where an Enertech 44/40 KW windpower generator has been
in operation since this spring. This installation, located at
Hull High School, is expected to generate 75,000 KWH this year,
saving the High School up to $7500. The turbine is horizontal
axis, downwind, free yaw; the rotor is 3-bladed, wood/epoxy
composite, 44 foot diameter; the generator is induction, 480 V,
3 phase, 60 Hz. Its rated power is 40 KW in 30 miles of wind.
Its tower is 80 feet high, galvanized steel, free standing
space truss. Of the $80,000 total project cost, $75,000 was
funded by the Massachusetts' Executive Office of Energy Resources.[7]

   CONCLUSION

It is possible and practical to establish a reliable alternative energy system which will demonstrate to the public that
basic energy needs can be met from renewable resources with
minimal reliance on diesel generators and without access to the
electrical grid.

The obstacles to implementation of such systems have more to do
with politics than with engineering, production or finances. In
order for political obstacles to be overcome, the public must
be educated.

---

[7] Energy Resources Newsletter   July/August 1985

The Use of an Electronic Maximum Power Point Tracker in Photovoltaic and Mixed Systems

J. M. M. Welschen

R & S Renewable Energy Systems BV, Eindhoven, The Netherlands

ABSTRACT

The use of a Maximum Power Point Tracker (MPPT) in photovoltaic systems to improve overall efficiency is investigated by means of computer simulation and practical experiments. Various types of control functions are used under different climatic conditions in small and large battery charging and waterpump systems. The results give a good indication when and where to use a MPPT and when not.

KEYWORDS

Photovoltaic; waterpumping; wind-photovoltaic hybrid systems.

INTRODUCTION

WHAT IS A MPPT AND WHY SHOULD YOU NEED ONE?

The electrical characteristic of a photovoltaic solar module is characterised by limited voltage and limited current capabilities (see fig.1). Therefore, the power which can be drawn is limited to a certain maximum. This maximum power can only be extracted at a particular voltage and current. The voltage at maximum power, Vmp, is mainly temperature dependent, decreasing with increasing temperature while the maximum power current depends mainly on the insolation, increasing with higher illumination level (see fig.2). Therefore, the maximum power that can be used from a solar module depends on the actual temperature and insolation.

Fig.3 indicates the area of maximum power points with all temperature and insolation levels. It is clear that there is hardly any load that will match such variety of operating points.

The most common situation is a battery charging system and because the battery voltage is almost constant under most charge current conditions, that voltage will dictate the solar module operating voltage.

B3.1

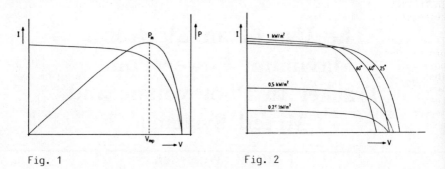

Fig. 1   Fig. 2

Designing a solar module for this purpose is a matter of compromise. As can
be seen in the power graph of fig.3, the slope on the right hand side of Vmp
is very steep and if Vmp shifts to the left with rising temperature the power
in the operating point will drop dramatically. On the other hand a larger
safety margin means more cells and therefore a more expensive module. If one
wants to standardise in module configurations also, an almost impossible
solution is asked.

A possible solution is a device that under all circumstances adapts the
battery voltage to the module maximum power voltage and in that way extracts
the maximum available power from the solar module.   Such a device is called a
Maximum Power Point Tracker, MPPT.

VARIOUS MPPT CONTROL CONFIGURATIONS

The power part of a MPPT is in fact a dc-dc converter with a transfer-ratio
set by the control part.   Since a MPPT has to optimise the solar module's
power, the control circuit has to multiply the current and voltage to obtain
the power and maximise that value by changing transfer-ratio.

An electronic continuous multiplier is rather complicated (compared to a simpl
straight forward solar system) and hence vulnerable and expensive. It is wort
looking for other, simpler solutions.

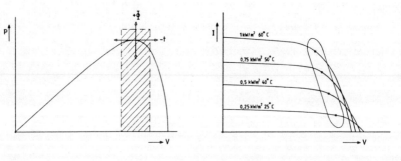

Fig. 3   Fig. 4

As indicated in fig.3 all possible maximum power points vary over a wide area
in the I-V field.   But it is very unlikely having low insolation and high
temperature or high insolation and low temperature.

If we assume a fixed relation between temperature and insolation, all
practical maximum power points are on a straight line (see fig.4).   The
position is determined either by measuring the temperature or the insolation.
That means that a conversion ratio, depending only on temperature or insolation,
must load the solar module with almost the maximum power voltage.   If we say
temperature that means cell temperature, but a further simplification is
measuring ambient temperature and taking NOCT into account.   Controlling in
accordance with insolation can be done by measuring the module current.

PERFORMANCE COMPARISON, SIMULATION RESULTS

For performance evaluation the following control characteristics are investigated:

    ID   a (theoretical) ideal MPPT with multiplier
    TP   control to a solar module temperature dependent voltage
    TO   control to an ambient temperature dependent voltage
    SD   control to a module current dependent voltage.
For reference:
    VPC   control to a fixed voltage.
    VA   direct connections to the battery

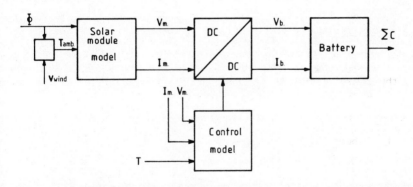

Fig. 5

To overcome the high costs of many field tests, a computer simulation on these
six systems was done first.   The following model was used (see fig.5).   As
input signals, standard days of various climatic types were used with insolation
and ambient temperature as function of time.   In some cases ambient temperature was calculated from insolation and windspeed.   As solar module model
an equation was developed which describes the V-I plane with 3-5% accuracy.
The various control functions were used on time step bases to calculate the
next steady state.   As battery model a large capacitor was used, giving a
reasonable voltage-state of charge characteristic.

Starting at 50% state of charge, the daily sum of added charge is the
performance figure.

74   J. M. M. Welschen

Table 1 gives the relative results of the different control functions for six
different climate types:
tropical with windspeed 1 m/s (trop.1)
tropical with windspeed 8 m/s (trop.8)
average with windspeed   1 m/s (av.1)
average with windspeed   8 m/s (av.8)
desert climate   (des)
winter   (win)

The reference value is the direct coupling to the battery (VA).

Percentage energy gain:

Control type	Climate					
	trop. 1	trop. 8	av. 1	av. 8	des	win
ID	3.2	4.2	6.9	12.3	22.9	31.6
TP	2.6	3.7	6.0	11.2		
TO	1.9	3.8	4.9	10.0	22.3	29.9
SD	- 4.2	- 7.3	- 3.3	- 2.8		
VPC	- 2.4	3.1	5.8	7.9		
VA	0	0	0	0	0	0

TABLE 1

As can be seen, both temperature dependent control functions are almost as
good as the ideal situation but only in extreme situations can a satisfactory
gain be achieved.

Fig.6 gives the anticipated   module voltage during the day in the tropical,
1 m/s windspeed situation. The ID curve is the exact momentane maximum
power voltage. TP and TO are optimised for minimum average deviation.   It
is remarkable that the module current control function (SD) gives such error.
This can be explained from the interaction between current deviation caused
by insolation variations and caused by converter ratio variation.

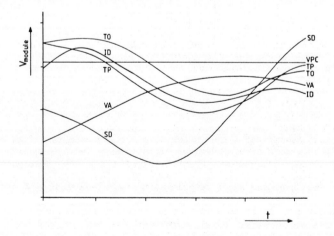

Fig. 6

PERFORMANCE VERSUS PRICE

The percentages in Table 1 are theoretical figures, taking not into account
possible mismatch or loss, in practical circuits. That means that in
practice, 10-25% gain might be expected in extreme situations (extreme hot or
extreme cold). In all other situations the gain is neglectable or even a
loss might be expected.

Fig.7 gives a price indication in US $ for various sizes of MPPTs. Depending
on control function and quality level, a tolerance of ± 25% has been
indicated.

Because a MPPT can also take care of the battery charge regulation, the cost
of a comparable regulator is subtracted.

Another solution to improve system output is adding modules. For a module
price of 6 $/Wp a break even point MPPT versus extra modules can only be
expected at 10% power gain in larger systems (see fig.7).

MPPTS IN LARGE PV AND HYBRID SYSTEMS

In larger solar or hybrid battery charging systems (>10 kWp) a considerable
higher system voltage has to be chosen (100-400 Vdc). The number of modules
in one series-string can range from ± 7 to ± 25. Even with standard solar
modules a good match can be looked for by adjusting the number of modules
in one string. Because of the large array area a great difference in insolation
and temperature must be expected (clouds, wind turbulence, partial shading,
etc.). Therefore, it is very difficult to predict the momentary maximum
power point per string, which is certainly not just the sum of the MPPs of the
modules.

More strings are connected in parallel, even combined with other power sources
(wind turbine, diesel, etc.). The battery acts as a voltage source and
hence all other sources must show a current source behaviour for system
stability. A string of solar modules can only be a "current source" when
operated left of the MPP. This means adding more modules to a strinn to be
on the safe side which is even more important when allowing a partial
module failure in the future. In such situations a combined device, MPPT
and "electronic current source" might be more economical than just oversizing
the array. The rapid decreasing costs per Wp for larger electronic units
will help to turn over the balance.

COMBINED PV-WIND INSTALLATION AT TERSCHELLING

The problems to find the right input signals for a computer simulation of
such a large system, combined with the complex simulation equations were the
reasons to start phase II and do practical field tests. The pilot plant at
Terschelling, sponsored by the CEC and local governments, was selected to
investigate the system behaviour under various conditions. Some key figures
of the plant are:
    50 kWp solar array
    40 kW wind turbine
    solar array divided into 29 strings
    one MPPT per string, specifications: U_{in} = 210 - 273 V
I_{in} = 0 - 7 A
U_{out}= 324 -432 V
I_{out}= 0 - 5 A

B3.5

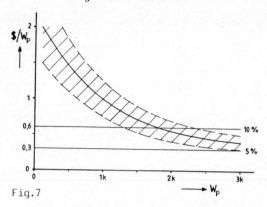

Fig.7

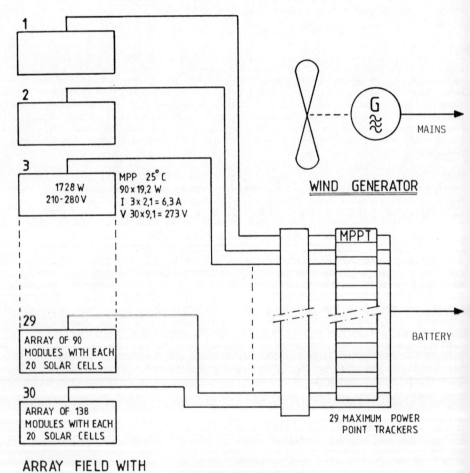

ARRAY FIELD WITH
30 ARRAYS

SOLAR GENERATOR

Fig.8

The control circuit of the MPPT is equipped with a micro-processor to enable
easy (software) modifications for different experimental control functions.
First setting is true multiplying on a time interval basis, stepping along
the I-V curve around the maximum power point.   Other control functions are
planned to be incorporated later. Fig.8 gives an overview over the generator
part.
Details can be found in (2) and (3).

Practical data and evaluation of results are not yet available but will be
published in the CEC result reports.

MPPTS IN WATERPUMP SYSTEMS

In a centrifugal water pump the impeller speed depends strongly on the depth
of the well. The motor voltage is proportional to the speed. Often it is
unknown what level has to be pumped and even if this is known, this may vary
quite often.   Therefore, it is not possible to predict at what voltage the
solar module has to operate and a considerable mismatch has to be expected.
A MPPT can adapt the module and waterpump characteristics.

To investigate the practical behaviour of such a system the waterpump project
at the hospital in Gao, Mali, was chosen for experiment. A 1.5 kW pump is
supplied by 16 strings of 2 modules each connected for 20 V and 2.5 A
typically.   The output voltage may vary from 20 - 60 V independent of the
momentary maximum power voltage either to a fixed adjustable voltage or an
ambient temperature dependent voltage.

All results of the theoretical analysis were confirmed, including the relative
high cost price for all MPPTs.   This is the main reason we switched over to
waterpumping systems with one MPPT (approx. 1.5 kW) at higher voltage for
efficiency improvement and with a temperature dependent voltage as control
reference.   Under worst case conditions an improvement of 50% compared to
direct module-motor connection is achievable.

CONCLUSIONS

With present price levels for solar modules and electronic devices, it is not
economically feasible to use MPPTs in small battery charging systems.   Only
in extreme temperature situations can a reasonable gain be achieved on
standard modules. In larger battery charging systems ($\sim$ 10 kWp) or systems
with unpredictable working points a MPPT can be a very valuable device, even
with the simplified control methods.

ACKNOWLEDGEMENT

The author wishes to thank Ir. R.Burgel, Ir. R.Sonneville, Ir. R.de Vries
and Ing G. Koelemeyer for their contribution to the research and projects on
which this paper is based.

REFERENCES

(1)   Ir.J.R. de Vries.   Ontwerp van een aanpassingeenheid voor een foto-
        voltaisch zonne-energiesysteem m.b.v. CAD technieken. May 17,1983
        report 1224.0416 University of Technology, Twente, the Netherlands.
(2)   Ir.R. Burgel, Ir. R.Sonneville, and Ir.J. Welschen. Optimization
        research into a complete photovoltaic generator/consumer appliance
        system for small independent electricity supply. Status report
        ESC-R-049-NR R & S Renewable Energy Systems BV, Eindhoven,

the Netherlands.
(3)   Ing.G. Koelemeyer. PV-Pilot Project Terschelling, the Netherlands.
      Holec Projects, Hengelo, the Netherlands.

TOPIC C
Biofuel

Saving Heating Oil on a Finnish Farm

E. H. Oksanen

Work Efficiency Association, Box 28, SF–00211 Helsinki,
Finland

ABSTRACT

Fuel oil consumption was cut down by installing a boiler using
wood chips as fuel and by installing a simple solar collector
for the cold air grain drier and chip drier.

KEYWORDS

Energy conservation, domestic fuel, wood energy, boilers,
solar collector, energy costs, wood chips.

INTRODUCTION

The annual energy consumption of Finland's agriculture,
including farmhouses and greenhouses, is about two million
equivalent oil tons. This is about 8 % of the total energy
consumption of the country. However, close to one fifth of the
light fuel oil consumed is used by agriculture to run diesel
engined machines and for heating grain driers and buildings.
Due to the sharp increase in oil prices in 1974-80, wood has
replaced heating oil on many farms which changed from wood to
oil in the 1960's and -70's. However, the majority of Finnish
farms have always relied on wood for heating of buildings.

The old wood heated stoves and boilers burned split billets or
one meter logs. This required much hand work in making the
billets and also in heating. Therefore most farmers changing
back to wood heating as well as those renewing their boilers
have selected heating with wood chips. Making chips is faster
and handling can be done with machines (Fig. 1). There was a
great rush to wood chip boilers at the end of the 1970's, when
many makes were still prototypes. The market levelled off in
1982, and the present day boilers are quite reliable and
require little attention.

THE CASE OF FARMER RAIMO

Raimo is farming 43 hectares of arable land and 90 hectares of
forest in South Finland. He is specialized in sugarbeet and
small grain production. He had plans to alter the dairy barn

C1.1

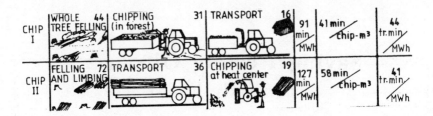

Fig. 1. Work methods and work input in making wood chips for
        farm purposes (Orava and others, 1983).

to a swine house and to keep two batches of 170 pigs for meat
during winter months. The main farm dwelling was built in 1902
and is to be repaired in 1986. Earlier the farm house was
heated with woodstoves burning split billets. The consumption
was about 50 m^3 of logs per year. In 1967 he installed an oil
heated boiler in the basement of the farm house. The oil
consumption was then about 7000 liters per year.

Continuously climbing oil prices made Raimo consider returning
to wood heating. His farm became one of the experimental sites
in a study by the Work Efficiency Association in 1981-83, in
which central heating plants for domestic combustible
material were constructed (Nurmisto and others, 1983). Since
1984, the farm has been the Finnish experimental site for the
FAO/UNDP Regional Project on Integrated Farm Energy Systems.

Besides the farm house, also the machinery repair shop at one
end of the dairy barn and the swine house at the other end
needed heating (Fig. 2). A cold air grain drier was
constructed in the attic of the barn. Heat exchangers, made
from a cooler of an old lorry and heated with hot water from
the chip heater boiler, were put in front of the blowers.
While building the driers, part of the wall panels in the
extension of the barn were made of perforated black metal
sheets. The drying air is first sucked in through this simple
solar collector, then through a heat exchanger. These two
together have raised the air temperature by 2-4 degree C and
speeded up the drying.

The Plan and its Realization

The heating needed during the winter season was calculated as
follows:

Buildings to be heated	Max. capacity needed	Heat consumption
Farm house	19 kW	49 000 kWh/year
Repair shop	(7 ")[1]	6 000 "
Swine house	6 "	3 500 "
	25 kW	58 500 kWh/year

1) Used only occasionally.

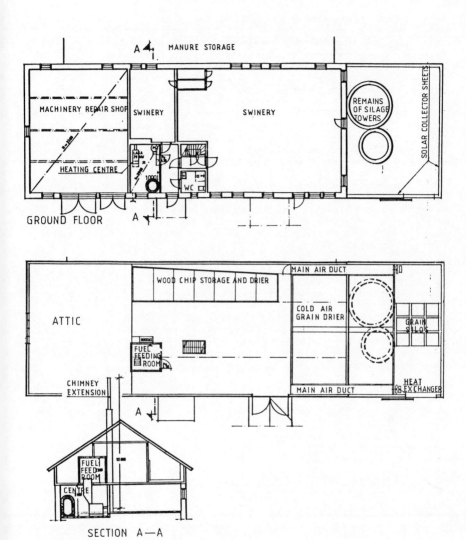

Fig. 2. The reconstructed barn of farmer Raimo.

The heat exchangers for the cold air grain drier get the full
effect of the boiler during the drying season in
August–September. The heating centre was constructed in the
middle part of the barn (Fig. 2).

C1.3

In this way the existing building space and chimney could be
utilized. It was favourable also from the viewpoint of
taxation. If the heating centre had been constructed in the
farm house, much less of the costs would have been deductible
from farm income in taxation. The traffic, noise, dust and
smoke caused by heating with wood chips would also have been
disturbing. The boiler selected for the heating centre was of
underburning type with a nominal capacity of 29 kW.

The wood chip for the boiler is made from waste wood, felling
residues and forest thinnings. Trees are felled in summer,
limbed and placed in small piles. The chipping is done the
next spring in the forest and chips transported with a tractor
trailer to the attic of the barn. The method is thus a mixture
of the methods featured in Fig. 1, with work input of two man
hours and 3/4 tractor hours per MWh, or a little less than one
man hour per chip-m^3. As thinning of the forest and removal of
waste wood from field outskirts must be done anyway, all the
work cannot be charged to the heating, due to felling and
limbing of chip trees.

Farmer Raimo, being a handy man, did almost all the
construction and building work himself during breaks in field
work. An electrician and a plumber were hired for their
special jobs. The farmer himself worked with the heating
centre 510 hrs and with the heating duct (56 m) from the
centre to the farm house 108 hrs. Getting the building
material, boiler etc. and organizing the work took 55 hrs,
totalling 673 hrs of work. Plumbing took about 100 hrs and
electrical work a couple of days.

The building costs, including his own work and materials from
the farm, were FIM 69,430,- (1 £ = 8,2 FIM). This was about
15 % more than calculated. However, as the value of his own
work and material were FIM 15 000,- or 22 % from the total
cost, the money expenditure was FIM 54 500,-. The state energy
subsidy was about 20 % of this or FIM 11 700,-. Thus the
investment made was FIM 42 800,-. A state farm loan of
FIM 44 000,- was given with low interest, so Raimo did not
have to invest his own funds for the purpose at all.

First Experiences

Raimo's heating centre was completed in January 1982 but could
be used from November -81 on. That winter was more or less a
learning period. The first batch of chips (63 m^3) was made in
August 1981 and the second (42 m^3) in January -82. Some
building wood residue was also burned. The moisture of the
chips when burned was 25-36 %, which proved to be too high,
causing trouble in heating and excessive consumption of chips.
Additional disturbances were due to dirt in pipings, sticking
valves, etc. The old oil boiler, kept in reserve was
occasionally put to short time use that winter.

Learning correct regulation of the chips boiler took quite a
long time. Keeping the boiler grate slots unclogged required
cleaning once or twice a day. The chimney also needed frequent

sweeping as it was slightly underdimensioned for the purpose.
Starting the fire in a cold boiler was slow. If the chips were
sticky or very moist, there was a tendency to vaulting in the
fuel duct. A few times the fire crept into the fuel duct.
Fortunately, no conflagration occurred, but minor explosions of
wood gases and plenty of smoke were common when filling the
fuel duct, specially when using building wood residues.

According to measurements, in the winter season 1981-82 Raimo
used 104 m^3 of chips with a heat value of 70 000 kWh. The
amount of building wood residue is counted as extra. Thus the
heat needed was greater than calculated, due to a hard winter
and the poor condition of the farm house. As the chips were
quite moist, some mould fungi occurred.   However, no illnesses
occurred due to it. The heating and cleaning work took on
average 30 minutes a day.

Later Experiences and Improvements

Due to trouble with moist chips, Raimo decided to build a cold
air chip drier in his chip store in the summer of 1982. This
has worked well. In June - July it takes only a few days to
dry a batch of 80-90 m^3 of chips from above 30 % to about 20 %
moisture. Even then a slight kserofil mould has appeared, but
it was present in the limbed trees already before chipping.
Raimo has not used the chip boiler in the summer. Hot water
for consumption has been heated by the old oil boiler. There
is a woodstove (range) in the kitchen, used year round, and
especially in the fall before heating with the new chip boiler has
been started.

During the second winter season, the conversion efficiency was
about 5 percent units higher than during the first one,
exceeding 60 % when loading of the boiler was over 50 %. The
rise was due to drier chips and better knowledge of the
regulation of the boiler. Very few disturbances occurred.
Smoking and explosions were rare, too. When loading of the
boiler was under 25 %, the conversion efficiency was about
45 %. The regulation of the boiler was then difficult.

In January 1985 Raimo put a stoker screw and a burning bowl
(head) on his boiler, which is no longer an underburning one.
However, the old fuel duct forms part of the chip store for
the boiler. This action has lessened the heating and cleaning
work considerably. At the same time the conversion efficiency
has been raised by about 7 %-units, from the earlier average
of 55 to 62 % (Tuomi, 1985).

During the winter season 1984-85 Raimo used 174 m^3 of chips.
The moisture was 19 % and the heat value 660 kWh/m^3. The chips
were from waste wood over a year old with some rotten and
mouldgd material in it. Had it been first class birch chips,
135 m^3 would have been enough. During the extremely cold
season in January - February 1985, about one m^3 of chips was
consumed daily. From the yearly amount of 174 m^3, about 25 m^3
was used for additional heat to the cold air grain drier and
about 20 m^3 for heating the repair shop and swinery.

Raimo is satisfied with his system now. Replacing some 11 500
liters of heating oil with 174 m^3 of chips seems sensible. The
average yearly heat production costs were calculated   in 1982
as follows:

Fuel needed:	150 m^3 of chip,	FIM 6 900,-
Heating work:	100 hrs	" 1 600,-
Electricity:		" 100,-
Maintenance and repairs:		" 500,-
Construction and equipment		" 4 000,-

		FIM 13 100,-

As the amount of heat produced with this cost is 53 000 kWh,
the cost per kWh is FIM 0,25. This is over 20 % less than the
corresponding figure for oil heating.

When asked about further actions to save heating oil on his
farm, Raimo names none. He sees hardly any use for biogas,
even if a biogas plant were possible on the farm. From what
his father told about the tractors run on wood producer gas
during and after the war, he is not interested in changing his
diesel tractors or the combine to run on a mixture of wood gas
and diesel fuel. Those are emergency solutions.

In 1986 he will repair the farm house and probably take out
the old oil boiler. The warm water needed in summertime will
then be produced by the wood chip stoker-boiler, which thus
will be in year round use.

ACKNOWLEDGEMENT

The author thanks farmer Raimo and the personnel of the Work
Efficiency Association named in references, who under the
author's leadership collected the study material and wrote the
main part of the handbook and leaflet. Grants for the study
were given by the Fund for Development of Agriculture.

REFERENCES

Nurmisto, U., S. Tuomi, A. Alanko, K. Turkkila and E. H.
         Oksanen (1983). Maatilan lämpökeskus kotimaiselle
         polttoaineelle. (Central heating plant of farm for
         domestic combustible material). Work Efficiency Ass.
         Publ. 254, 1-181.
Orava, R., S. Ryynänen, A. Siekkinen and H. Solmio (1983).
         Työnkäyttö kotoisen polttoaineen korjuussa.
         (Consumption of work when harvesting domestic fuel).
         Work Efficiency Ass. Agric. Rep. 6/1983, 1-6.
Tuomi, S. (1985). Kokemuksia hakelämmityksestä FAO:n energia-
         verkostoprojektin Suomen seurantatilalta.
         (Experiences in wood chip heating at the Finnish
         experimental site of FAO's energy network project).
         Accepted for publication.

Marine Biofuel for Rural Coastal and Island Communities in Developing Nations

B. D. Gold and E. B. Shultz, Jr.

Department of Engineering and Policy, School of
Engineering and Applied Science, Washington University,
St. Louis, MO 63130, USA

ABSTRACT

Certain seaweeds may be grown in coastal waters for conversion to biogas
to supply energy for integrated development of rural coastal centers of
production.

KEYWORDS

Seaweeds; *Gracilaria*; *Eucheuma*; biogas; anaerobic digestion; integrated
development; rural employment generation.

INTRODUCTION

Two crucial problems facing rural communities in developing nations are
the needs for adequate fuel and economic development to stem the migration
of people to the cities. In this paper we discuss an "integrated coastal
village system" (ICVS) as a possible solution to these problems for rural
coastal and island communities. We first referred to such a scheme centered
around seaweed farming for multiple uses in 1983 (Gold and Shultz); the
present paper develops the concept further.

Swaminathan (1980) described a similar approach but did not include the
potential for energy independence. He pointed out that development of
marine biomass resources might form the basis of integrated coastal development
providing income and employment for the rural poor. His multi-faceted
design concept called for capture and culture fishery activities combined
with silviculture at the water's edge for additional cash crops. Because
the provision of fuel was not mentioned, we believe that Swaminathan's
model needs to be expanded to include energy independence in rural coastal
and island communities; importation of fuel from urban to rural centers
or extension of the electricity grid are likely to be costly alternatives.

Included in the ICVS would be fish and shellfish harbored by seaweeds,
conversion of seaweed to biogas, use of biogas for home cooking and lighting

fuel, conversion of some biogas to electricity, development of small-scale
industries using biogas and electrical energy to process locally-produced
materials from agriculture, silviculture, and mariculture (e.g. phycocolloids
and shipping of products to urban markets using biogas-derived compressed
methane or locally-produced alcohol as fuel for boats or trucks. As described
in our 1983 paper, the ICVS is integrated at two levels: both integration
within the technical elements that internally comprise the system; and
externally toward the type of support and infrastructure and local involvement
in decision-making required for such a project to be successful.

Revelle (1978) identified the crux of the rural employment problem:

> With the small and decreasing size of agricultural land-holdings
> per farm family and the growing numbers of rural people without
> land, it will probably be impossible to raise rural average
> incomes to a satisfactory level unless employment can be increased
> through the development of small industries in the countryside.

He further argued that to be competitive with industry in large cities,
rural small-scale industry must have inexpensive energy.

Approximately 70 percent of the people of developing nations live in rural
areas and rely on non-commercial fuels (Baxendell, 1981). As a result,
firewood, agricultural crop residues, and animal dung have been over-exploited.
Therefore, in those regions where seaweed grows or can be grown, it may
provide an attractive alternative to terrestrial biomass. Simple coastal
seaweed farming techniques have been developed in the Philippines (Edwards,
1979) to produce *Eucheuma* for its carageenan, a food colloid. In 1974,
over 10,000 metric tons of *Eucheuma* were exported from Tanzania (Mshigeni,
1979). In a similar example, *Gracilaria* has been farmed in Taiwan in seawater
ponds (Shang, 1976).

EXTENSION OF THE ICVS CONCEPT

In addition to conventional agriculturally-based industries, novel projects
might also be incorporated into the ICVS concept. For example, agricultural,
silvicultural or maricultural materials might be locally processed to
semi-finished products before shipping to urban areas for further processing
by capital-intensive methods. This "semi-processing" approach (Schuplin
and co-workers, 1981) would add value and increase income to the rural
people, without requiring expensive advanced technologies to refine the
semi-finished product.

An example is the extraction of oil from novel semi-tropical oilseeds
such as the Chinese tallow tree (*Sapium sebiferum*). This would provide
a source of highly-valued industrial chemicals for plasticizer, synthetic
lubricant and speciality polymer markets. The processing could be carried
as far as the stage of a crude stabilized oil at the ICVS level using
relatively simple, capital-conservative equipment. At this point, the
oil could be transported to urban centers for refining to final products
in sophisticated equipment at large scale (Scheld and co-workers, 1984).

REVIEW AND ANALYSIS OF SEAWEED FARMING EXPERIENCE

Seaweeds are promising for mariculture because they are primary producers
with high growth rates, can be a renewable source of biomass for conversion

to biogas, and yield valuable industrial and food products. Ultimately, the importance of seaweed farms in providing energy for rural coastal and island communities will depend on their productivity. Factors which can affect productivity include: choice of species; nutrient levels, especially inorganic nitrogen; solar radiation; water temperature, turbulence and turbidity; frequency and magnitude of storms; grazing; competition; and disease.

Most coasts have waters that are relatively high in nutrients supplied by either natural or anthropogenic sources. Tropical coasts with high light incidence and high temperature throughout the year should be able to support year-round seaweed farming. Jackson and North (1973) and Jackson (1981) have conducted detailed reviews of the factors affecting seaweed production, the suitability of species for mariculture, and criteria for site selection.

Seaweeds may be farmed by "extensive" or "open" methods in natural bodies of water or by "intensive" or "closed" methods in ponds, tanks, or raceways. In Asia, warm-water seaweeds such as *Gracilaria* and *Eucheuma* have been successfully cultivated using simple techniques on a small scale, for production of food substances. Japanese, Chinese, Taiwanese, and Philippine experiences, among others, have been reviewed (Gold and Shultz, 1983).

In the USA, attention to seaweed farming occurred initially in California (Wilcox, 1980) based on *Macrocystis pyrifera*. More recently under the Department of Energy and the Gas Research Institute marine biomass program, research is being conducted in New York and Florida, and *Gracilaria* and *Laminaria* are included (GRI, 1982). This program recognizes the considerable promise of biogas from marine biomass as a contribution to the energy mix of the United States. Ryther, et al., (1978), and Ryther (1981) have screened over 40 species of seaweeds. The red alga *Gracilaria tikvahiae* had the highest sustained yields. Ryther, et al., (1978) conservatively estimated potential yields of 32.4 MT/ha-yr (dry weight). Brinkhuis and Hanisak (1981) have screened nine species of red, brown and green seaweeds as candidates for biomass production in Long Island Sound, and subsequent conversion to methane.

Yield information available for an original planting of 200 g of *Eucheuma* in a Philippine coastal seaweed farm indicates that a farm typically produces 3 kg in three months, a 15 to 1 increase in weight, and this rate of production can continue on a year-round basis. A typical 0.5 hectare Philippine seaweed farm will yield approximately 10-15 MT/ha-yr (dry weight) (Edwards, 1979). Furthermore, one family (typically four persons) can operate a farm this size and still carry out the usual subsistence activities (Edwards, 1977).

REVIEW AND ANALYSIS OF SEAWEED BIOGASIFICATION EXPERIENCE

Anaerobic digestion of seaweed has been demonstrated; it is not hampered by the salt water on the seaweed (Ghosh, et al., 1976; Klass, et al., 1978; Chynoweth, et al., 1981; Hanisak, 1981a). The potential of low-energy biogasification (minimal pre-processing, no energy for heating and stirring) has also been shown (Hanisak, 1981a). Fannin and co-workers (1983), have demonstrated the advantages of upflow digestion of seaweeds in their upflow solids reactor (USR). This digester design seems to be simple and effective; biogas yields ranged from 0.62 to 0.66 standard cubic meters (SCM) per

kg of volatile solids (VS), over a wide range of loading (throughput rates).
Biogas from digestion of the seaweed *Macrocystis pyrifera* in the upflow solids
reactor contains about 56 percent methane (Fannin and co-workers, 1983),
the remainder being carbon dioxide with small amounts of other gases.
The heating value is about 60 percent of that of methane.

Hanisak (1981b) has proposed the possibility of extracting agar and methane
from cultured *Gracilaria* and recycling the nitrogen from the digester residue
back to the seaweed farm in an integrated utilization scheme. He has
demonstrated the feasibility of recycling 73 percent of the nitrogen from
the digester back to the cultures.

CALCULATION OF POTENTIAL BIOGAS YIELD FROM SEAWEED FARMING

It is of interest to estimate the potential yield of biogas from a half-
hectare family (four person) farm, growing *Gracilaria* or *Eucheuma*. Average
dry seaweed yields of these species appear to be similar, about 16 MT/
ha-yr. For a half-hectare this would be 8 MT (8,000 kg) of dry seaweed
annually. Fannin et al., (1983) reported an average yield of 0.64 SCM
biogas per kg of volatile solids (VS) from *Macrocystis* in the upflow solids
reactor, and *Gracilaria* contains about 58 percent VS (Hanisak, 1981a).
Multiplication of these three numbers leads to an estimate of 3,000 SCM
biogas per half-hectare per year. We expect that results would not be
much different if *Gracilaria* were biogasified in the same reactor. Hanisak
(1981a) has observed that digestion of *Gracilaria* and *Macrocystis* proceed
similarly at long retention times. Further, digestion of *Eucheuma* should
also be similar. Both *Gracilaria* and *Eucheuma* are red seaweeds with high
percentages of phycocolloids which are similar in chemical structure.

Our estimated annual production of 3,000 SCM of biogas per family may
be compared with an estimated 365 SCM required annually by a Thai family
of four, for its cooking needs (Biogas Newsletter, 1980). Family biogas
requirements for lighting will depend on numbers of lamps and mantles,
but can be estimated as roughly equal to biogas requirements for cooking
fuel, based on information given by Sathianathan (1978).

Therefore, it appears that labor-intensive family farming of seaweed on
onehalf hectare plots in Southeast Asia has the potential of producing
about four times as much biogas as the family needs for cooking plus lighting
(730 SCM). The excess, about 2,300 SCM annually per family, would be
available for community needs and local industries.

ASSESSMENT AND IMPLEMENTATION OF THE CONCEPT

An assessment of the potential for seaweed farms to meet energy needs
of developing nations must take into account these factors, among others:
the extent of natural stocks for direct harvest or source of seedstock;
potential yields of biomass; conversion efficiencies; economics; environmental
impacts; institutional and infrastructure requirements; and sociocultural
considerations (Gold and Shultz, 1983).

Some important implementation questions include: is the village size
and organization appropriate to the ICVS scheme; what types of changes,
if any, will come about from the implementation of the ICVS scheme; is
the infrastructure necessary to support an ICVS available in the village

or from a nearby center; what is the relation between the village and
its closest urban center; does the coastal site adequately supply the
alga's physical and biological requirements; what competition with other
existing or anticipated uses must be reconciled; what is the relation
of the site to pollution sources (adverse or beneficial); is there enough
space available for a viable ICVS and to provide room for expansion; what
legal and political constraints exist; are there adverse public attitudes
and perceptions to the ICVS scheme; is the available labor (skilled and
unskilled) amenable to the working conditions of the marine environment;
and will appropriate research facilities and extension services be readily
accessible (Gold and Shultz, 1983).

Experience with rural development projects generally indicates that one
of the keys to the success of a project is its acceptance by the local
people and their desire to see that it works. This generally requires
that the project not be imposed on the villagers, but that the concept
be introduced to them in such a way that they be involved in its planning
and development and that after successful demonstration it be turned over
to them.

CONCLUSIONS

Based on our analysis of the seaweed farming and biogasification literature,
and our calculation of potential yields of biogas from small-scale coastal
seaweed farming before and after household fuel needs are met, we conclude
that the ICVS scheme may be useful and that it deserves further evaluation.

REFERENCES

Baxendell, P. B. (1981). Development, 2, 61.
Biogas Newsletter (1980). No. 9. University Press, Tribhuvan University,
    Kathmandu, Nepal.
Brinkhuis, B. H., and M. D. Hanisak (1982). Development of a marine biomass
    program in New York. Proceedings 1981 International Gas Research Conference,
    Government Institutes, Inc., Rockville, MD.
Chynoweth, D. P., et al. (1981). Chem. Engng. Prog. 77, 48-58.
Edwards, P. (1977). Seaweed farms: an integral part of rural development
    in Asia, with special reference to Thailand. Rural Development Technology:
    An Integrated Approach, 665-680. Asian Institute of Technology, Bangkok,
    Thailand.
Edwards, P. (1979). Appropriate Technology, 6, No. 1, 25-27.
Fannin, K. F., V. J. Srivastava, D. P. Chynoweth, and K. T. Bird (1983).
    Energy from Biomass and Wastes, 7. Institute of Gas Technology, Chicago,
    IL, 563-584.
Gas Research Institute (1982). Research and Development Results - 1981.
    Gas Research Institute, Chicago, IL.
Ghosh, S., et al. (1976). Research Study to Determine the Feasibility
    of Producing Methane Gas from Sea Kelp. Institute of Gas Technology,
    Chicago, IL.
Gold, B. D., and E. B. Shultz, Jr. (1983). Integrated rural development
    for coastal villages: schemes based on biomass farming. Presented
    at 9th Annual Third World Conference, Chicago, IL, March 23-26.
Hanisak, M. D. (1981a). Methane production from the red seaweed
    Gracilaria tikvahiae. Proceedings Xth Int'l Seaweed Symposium, Walter de
    Guyter and Co., New York.

Hanisak, M. D. (1981b). Bot. Mar. 24, 57-61.
Jackson, G. A. (1981). Biological constraints on seaweed culture, in Biomass as a Nonfossil Fuel Source, D. L. Klass, ed., American Chemical Society, Washington, DC.
Jackson, G. A., and W. J. North (1973). Concerning the selection of seaweed suitable for mass cultivation in a number of "marine farms" in order to provide a source of organic matter for conversion to feed, synthetic fuels and electrical energy. Final Report to U.S. Naval Weapons Center, China Lake. Kech Laboratory, Calif. Inst. of Tech., Pasadena, CA.
Klass, D. L., Chynoweth, D. P., and S. Ghosh (1978). Biomethanation of giant brown kelp Macrocystis pyrifera, in Energy from Biomass and Wastes, Institute of Gas Technology, Chicago, IL.
Mshigeni, K. E. (1980). Rural Tech. Bull., Oct.-Dec. (No. 3) 22-26.
Revelle, R. (1978). Requirements for energy in the rural areas of developing countries, in Renewable Energy Resources and Rural Applications in the Developing World, N. L. Brown, ed., Westview Press, Boulder, CO.
Ryther, J. H. (1981). Proceedings 3rd Annual Biomass Energy Systems Conference, Solar Energy Research Institute, Golden, CO.
Ryther, J. H., J. A. Deboer, and B. E. Lapointe (1978). Cultivation of seaweeds for hydrocolloids, waste treatment, and biomass for energy conversion. Proceedings Int'l Seaweed Symposium 9, 1-16.
Sathianathan, M. A. (1978). Biogas achievements and challenges, in Biogas Technology in the Third World, Andrew Barnett et al., IDRC, Ottawa, Ontario, Canada, p. 63.
Scheld, H. W., et al. (1984). Seeds of the Chinese tallow tree as a source of chemicals and fuel, in Fuels and Chemicals from Oilseeds, E. B. Shultz, Jr., and R. P. Morgan, eds., Westview Press, Boulder, CO.
Schuplin, T. J., et al. (1982). Rural semi-processing of agricultural products: appropriate technology for oilseeds. Presented at 5th National Conference on the Third World, University of Nebraska at Omaha, Omaha, NE.
Shang, Y. C. (1976). Aquaculture 8, 1-7.
Swaminathan, M. S. (1980). Mazingira 4, (No. 1) 43.
Wilcox, H. A. (1980). Prospects for farming the open ocean, in Bioresources for Development: The Renewable Way of Life, A. King and H. Cleveland, eds., Pergamon Press, New York.

Peat and Energy Production

M. Herbst and D. Brickenden

Herbst Group of Companies, Head Office, Ireland

ABSTRACT

An integrated approach to peat production, distribution and conversion to useful energy is outlined. The components of this system are described as are the benefits. The principle could equally be applied to other forms of biomass.

KEYWORDS

Remote and island communities, unemployment, peat, energy, peatland, Herbst peat production system, Herbst Gasmiser, Producer Gas Generator, Milkman System, biomass, gasifier.

INTRODUCTION

Considering what appears to be a widespread exodus of the inhabitants of most remote and island communities towards the traditional and already overpopulated industrial/commercial centres, the problem of energy supply in these areas is likely to be short lived. With no inhabitants clearly there can be no problem. This scenario is outstanding with respect to such communities associated with the 'developed' nations and though somewhat less so, still apparent in the developing countries.

This erosion of our social system must not only be contained, but be reversed and the solution(s) must be capable of being implemented now and offer long term prospects.

There are two paths by which this problem may be approached. One is by the direct injection of capital in the form of social welfare and subsidies to allow subsistence survival. The more satisfactory path is to take advantage of the existing resources, namely the land and the people, and prepare a basis for the long term development of the community.

It is generally accepted that energy is a key factor to the survival of these communities and to substantially reduce the price that is currently being paid for that energy would in most cases, mitigate the circumstances.

Reference community

*   Agricultural and/or marine based economy with seasonal support from the
    tourist industry.
*   Subsistence lifestyle.
*   Economically deprived.
*   Politically weak (in that they would find it difficult to effect change
    in Government policy).
*   High rate of emigration of young population.

Listed above are the main factors analogous to most remote and island
communities. They clearly indicate that a solution must:

1.   ....be capable of rapid implementation.
2.   ....utilise technologies and processes which are readily available
             and proven and not alien to the population.
3.   ....contribute to a secure base for the continuing development of
             that community.

The questions that remain unanswered, the answers to which may be central to
the fact that a solution has not yet been implemented, are:

*   Why have the ideas which have proven economically and technically viable
    in these circumstances generally not gone beyond demonstration stage?

*   Why does it appear that the level of government interest and economic
    support for these projects is low?

It may not be appropriate to attempt to answer these questions within this
forum. However, by inference the answers will reflect the political nature
of decisions that are or will be taken with regard to the social, economic
or energy policies pertaining to the remote and island communities.

Obviously great care must be taken when making any decision which could bring
about change, that the result is not transient and debilitating. A situation
akin to that apparent in Scotland at the moment is not desirable.   That is,
the growing number of 'ghost' towns which are residual to the start up of the
oil industry.   Another salient point, again pertaining to the oil industry
most directly, but likewise effected by the other convenience fuel industries
is the following Catch 22 situation: The demise of many industries particularly those that are energy intensive has been a direct result of the
exorbitant cost of energy.   There is an escalating problem of unemployment.

The Governments concerned require the revenue from the convenience fuels to
subsidise the maintenance of the growing unemployed population.   The powerful oil and coal lobbies share this interest with the government.

The resultant high price for energy is bringing about the closure of many
companies thus increasing levels of unemployment and so on.

A Solution

With the technology available today   many areas, which are economically
deprived at present, could have the lowest energy price via indigenous peat
reserves and the production of biomass on marginal land.

It is possible to produce from one acre of peatland up to 44000 litres of oil
equivalent/annum at a production cost as low as 5.30 pounds/TOE or 2.50 per

C3.2

tonne peat 35%.   This is the lowest energy price available to the U.K. and
the resource is virtually untapped.   10% of the land area in Scotland is
peatland.   There are large peat deposits in Wales and other parts of the
country, yet there is no government policy or support for such an obvious
source of energy.   The peat industry is seriously hindered due to this lack
of overall policy, permission to exploit is delayed as no guidelines exist
for the administrators and a charge is levied on all peat produced.   The U.K.
is unique in the E.C. countries with this attitude to such potential.   In
countries like Ireland private sector involvement in peat exploitation is
encouraged with grant aid up to 60% of the cost of equipment and ground preparation.   Peat plays an important role in the economy of countries such as
Sweden, Finland, Ireland, Russia and there is no good reason why it should
not be so in the U.K.

The Herbst Peat Production System

Four years ago Herbst Peat and Energy, a member of the Herbst Group of
companies, identified an area which apparently had a very high potential in
energy terms and was largely unexploited.

The potential energy was in the form of peat lying in bogs which would have
been considered marginal or not commercially viable by the conventional industry
at that time.   Until recently peat production on a commercial scale was
carried out by companies, typically state or semi-state, on vast peatland
areas using large, cumbersome, dedicated machinery.

In general these 'marginal' bogs are located in remote districts and aside
from some handwon turf being produced they would have had little other use.

With this in mind Herbst Peat and Energy developed a system appropriate to
the circumstances.   The heart of this system is an agricultural tractor
(75 hp).   All of the attachments necessary for draining, cutting, turning,
harvesting and transporting the peat can be linked to this basic power unit.
The basic attachment is the peat cutter and with a standard tractor drawn unit
the output is 7-10 tonnes/hour.

Figure 1   Tractor mounted Herbst peat cutter

Figure 2   The Bogmiser, a self propelled unit, has an output up to 50 tonnes/
            hour. This unit can do all work on the bogland and saves road and
            bridge building.

The Herbst system was first introduced in 1981. In Ireland in 1984, one
million tonnes of peat were produced with this revolutionary peat cutting
system. In Burundi last year 20,000 tonnes of peat were produced. With the
exception of wood this is their only indigenous fuel. The system has proven
to be a viable alternative in Canada, U.S.A. and the Falkland Islands.

It is necessary to ensure that the production of peat itself is not seen as
the only key factor to employment generation. The greatest potential lies
in the efficient utilisation of this low cost energy form. It is important
not to fall into the temptation to briquette or synthesise unnecessarily thus
increasing cost of production when the original peat produced at low cost
satisfies most requirements.

Low cost energy attracts energy hungry industries and if used at point of
production it is hard to find energy at such a low price. This has been the
case in other energy sectors. Good examples are the Ruhrgebiet in Germany
and Sheffield in the U.K., both industrial centres which developed close to
the point of fuel production.

Peat for Energy

Historically in the U.K. peat has received little credit as a fuel in the
form in which it normally comes off the bog, that is 35-40% moisture by wet
weight, 350-450 kg/m^3, in sods 300-450 mm long with a net calorific value of
12.6 MJ/kg.

Fluidised bed and other large scale combustion units are on the market. Until
recently, however, efficient small to medium scale combustion systems suitable
for peat have not been available.

As an extension of the principle of low cost, reduced scale peat production
methods it was essential to identify a means of converting the peat to useful
energy which would be acceptable to a wide range of consumers and also require
a minimum of alteration to its cut form. A collaborative exercise was
initiated involving Herbst Peat and Energy and Herbst Germany. The exercise
involved extensive investigation of the technology of heat and power production from solid fuels, attitudes and requirements in the various sectors
(domestic, commercial, and industrial) and identification of available hardwar

The main results were as follows:

The most appropriate use for peat and other forms of biomass for energy purposes is heat production, especially where it can replace fuel oil.

A system based on the principle of partial gasification (volatilisation) should be developed for conversion of the fuel to useful energy.

Gasification of peat and other forms of biomass for power and heat production is of particular interest to developing countries and the state of the art in this technology is appropriate to these countries.

A volatiliser has been developed and is now being manufactured by Servotherm Ireland Ltd, a member of the Herbst Group of companies. It is marketed under the name Herbst Gasmiser. Outstanding features of this system are:

1. Suitable for retrofitting to most existing oil, gas or solid fuel fired boiler plant.
2. Efficient in operation - seasonal thermal efficiencies for complete systems (boiler, burner, distribution system and controls) of 70% are attainable.
3. A high degree of automation is possible:
   augur feeding from intermediate storage
   augur removal of ash
   vibramotors for grate shaking
   "   for preventing hopper bridging with certain fuels
4. A variety of fuels can be used including woodchips, woodcubes, sawdust, crushed sod peat. In most cases the only alterations required for switching fuel are a change of grate and adjustment of primary and secondary combustion air settings.
5. Crushed sod peat with a moisture content up to 45% by wet weight can be utilised efficiently.

Figure 3. Herbst Gasmiser 60 kW fitted to boiler

the main components of the Herbst Gasmiser are:
1.   The reactor chamber - this is a monolithic refractory mass surrounded
      by high density insulation. Located in the main chamber are a grate
      and grate bearers with cams and over the grate is the reactor.   All
      components in the high temperature zone are cast from a high grade
      iron alloy.
2.   The fuel hopper - this is a gas tight construction of mild or stainless
      steel (dependent on main fuel used).
3.   The combustion tunnel - again a monolithic casting from the same
      refractory material as the reactor chamber.

A series of units are manufactured in the range 30-600 kW and in certain circumstances multiples of these units can be installed.

Producer Gas Generator

This is a downdraft gasifier with an integrated gas cooling and scrubbing
train.   The system can be used with a wide variety of fuels and is especially
suitable for gasification of peat.   Generally it is supplied as part of a
total energy package including a modified diesel engine driving an asynchronous
generator.   Various options are available for heat recovery from the producer
gas engine and generator. The engine can be converted fully to producer gas
by lowering the compression ratio and installation of spark ignition.   The
end result is in effect a gas engine. Alternatively a standard diesel engine
with little alteration can be run in a dual fuel mode.   In this case about
85% of the engine power output is drawn from the producer gas, the remaining
diesel being necessary for ignition.

The standard unit can be matched with engines to give outputs up to 100 kW.
If higher output is required, multiple units are set up. This permits the
maintenance of the proven and low cost technology.

The Milkman System

The Milkman system is the logical extension of the principles referred to
throughout this paper. It was developed for the peat industry but can
equally be applied to biomass production especially energy forestry.

The purpose of this approach is to simulate the convenience of the convenience
fuels.

Components of the concept are:

1.   The producer/distributor of locally sourced rawfuel, i.e. an
      Energy Farmer/Energy Merchant supplies. installs and services the
      necessary combustion equipment and heat flow meters. He will
      guarantee fuel availability and supply of energy to the consumer at
      an assured discount from reference energy costs e.g. heating oil/
      natural gas.

2.   The consumer is charged for metered energy use over fixed periods
      in similar manner to utility energy producers.

Figure 4.   Milkman System

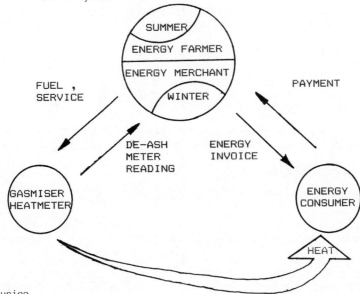

Conclusion

The solution outlined above is an immediate answer to the problem outstanding.
Once the peat is exhausted the remaining land (with 50 cm of peat left behind)
can be utilised for agriculture or the growing of biomass.   The production
of peat and biomass offers profit for all :   the landowners, the community,
the nation and the end user.   Government input is necessary to initiate
this solution.

Anaerobic Digestion of High Solids Feedstock

T. M. Clarke and A. J. Murcott

Farm Gas Ltd., Unit 2, Industrial Estate, Bishop's Castle,
Shropshire SY9 5AQ, UK

ABSTRACT

Practical experience with conventional fully mixed mesophilic
anaerobic digesters of 1.5 m3, 10 m3 and 105 m3 has
demonstrated the ability to process feedstocks containing more
than 15% total solids (T.S.) on a continuous basis producing
significant nett energy yields.

KEYWORDS

High solids feedstocks; continuous process; conventional
reactor; energy balance; maceration; separation.

INTRODUCTION

More than 75% of agricultural wastes in the U.K. may be
classified as 'high solids' containing 15% or more total
solids and a similar proportion occurs in other European
countries. For instance, in France 80% of farm wastes are
available as straw based manure (Finck & Goma, 1983).

The major limitation of anaerobic digestion as a source of
energy in the past has been its restricted application to
relatively homogeneous feedstocks containing generally less than
10% solids e.g. animal slurries, sewage sludge, industrial
wastes. For solids concentrations above 10% in the fully mixed
type of reactor increasing practical difficulty has usually
been experienced due to formation of a floating fibrous solids
layer inhibiting the digestion process both mechanically and
biochemically.

Attempts to overcome this problem have usually entailed costly
solutions involving pre-treatment, engineering, labour, energy
etc.

Addition of water to aid the mechanics of digestion adversely
affects the energy balance. Conversely, reduction of moisture

to approximately 25-30% T.S.   does not appear to significantly
reduce specific gas yield (Wujick, 1979; Jewell W.J., 1978).

RESULTS AND DISCUSSION

Three single stage fully mixed digesters of 1.5 m3, 10 m3 and
105 m3 capacity were installed to process a variety of mainly
high solids wastes.   The feedstocks were added to the
digesters, without pre-treatment, daily or on the basis of a 5
day working week.

Digester A   :   1.5 m3 experimental digester to assess
                    feasibility of treatment of abattoir wastes.
                    Feedstock :   three types ranging from 5-16%
                    total solids.

Digester B   :   10   m3   digester   on   isolated farm/
                    smallholding.
                    Feedstock   :   Straw   based   manure
                    approximately 18% total solids.
                    Gas   Utilisation :   Fuelling   "Rayburn"
                    cooker.
                    Digested waste was separated into solids
                    and liquid residues for more effective   use
                    on land.

Digester C   :   105   m3   demonstration   digester for
                    treatment of abattoir wastes.   Installed
                    for purpose of   waste treatment.
                    Feedstock :   four different types of waste
                    with total solids content of 5-20%+.
                    Gas Utilisation : Firing dual fuel high
                    pressure steam boiler.

Fig. 1.

Digester System C. showing digester (right),
gas bag (rear) and digested waste storage tank

With a mixture of different feedstocks some very dry organic wastes could be incorporated, but this type of digester would probably be limited to an average feedstock solids concentration of 25% maximum depending on digestibility. The upper limits, however, have yet to be defined.

Results obtained from Digester B indicate that a digester of 100 m3 capacity treating a typical high solids feedstock, such as straw based manure, would have the following performance :

TABLE 2 ESTIMATED ENERGY BALANCE OF 100 M3 DIGESTER

Feedstock Loading Rate	:	2 tonnes/day at 18% T.S.
Retention Time	:	50 days
Total Solids Loading Rate	:	360 kg/day
Specific Gas Yield	:	0.35 m3/kg T.S.
Gross Gas Yield	:	126 m3/day at 65% CH4
	=	3210 MJ/day
Gross Energy Input (Thermal and Mechanical)	:	480 MJ/day*
Nett Energy Output	:	2730.5 MJ/day

*Ambient Temperature : 10°C

On this basis two 70 m3 digesters (Farm Gas Model FGB70) are currently being constructed on small farms of less than 50 acres.

Fig. 2

Manual loading of Farm Yard Manure in Digester System B.

To further improve efficiency of digestion, a low powered (1.1 kW) macerator has been developed. Trials with digesters B and C suggest that maceration during digestion may increase specific gas yields and/or allow for increased loading rates with a very favourable effect on the net energy balance. Badger, Bogue and Stewart (1979) found that batch digestion of pre-macerated straw gave specific gas yields of 0.5 m3/kg T.S.

T. M. Clarke and A. J. Murcott

TABLE 1   SUMMARY OF RESULTS OF DEMONSTRATION DIGESTERS

Digester	A	B	C
Capacity (m3)	1.5	10	105
Period of Measurement (months)	5	5	3
Feedstock Type	abattoir wastes	straw based manure	abattoir wastes
Continuity of Loading (no. of days/week)	5	7	5
Feedstock % T.S.	5-16	18	5-20
Average Feedstock % T.S.	14	18	14
Digestion Temperature (°C)	35	35	35
Loading Rate (kg/day)	30	156	600
Gross Gas Yield (m3/day)	1.44	10.2	43
Specific Gas Yield m3/kg T.S.	0.34	0.36	0.40
*Average Methane Content (%CH4)	60	65	60
Gross Energy Yield MJ/day	40.0	259.9	1011.4
Combined mechanical and heat input MJ/day	24.1	47.9	292
Nett energy output MJ/day	15.9	212	719.4

*Estimated from intermittent analysis

The main objective of Digester C (currently in continuous operation) was to prove its effectiveness as a waste treatment system. Due to the experimental nature of the installation, loading rates were much lower than the capability of the system, and could be increased to approximately 4 tonnes per day, i.e. more than 4 times the present rate as the major constituent of the wastes (i.e. paunchings are highly digestible.)

During this period of monitoring, difficulties were experienced with the mechanical aspects of loading of dry straw based manure into the digester.

With increasing T.S. content of the digested waste it is increasingly desirable to separate the effluent into the high solids fibrous residues and a homogeneous liquid effluent containing, for instance, 3-4% T.S. Separation increases the ease of handling of the respective residues and makes the high solids fraction potentially saleable.

At present, high solids agricultural wastes are generally not treated prior to use or disposal. However, the anaerobic digester must be viewed in the context of effective recycling of wastes and conversion to a more useful form with energy as a major by-product.

Important design features of this type of digester include :
-Tanks constructed from G.R.P. with integral polyurethane foam insulation facilitating transport and installation and minimising heat loss.
-The ability to evenly mix digester contents and effectively transfer heat in a high solids medium.
-Simple and robust mechanisation requiring maintenance compatible with that of basic agricultural machinery (lubrication etc.).
-Fully automatic control of operation which also regulates digester start-up without skilled supervision. Labour is limited to supply of feedstock to the digester.

As a viable source of energy (especially for remote communities) digestion must satisfy, as far as possible, the following criteria.

a. Low cost
b. Ease of maintenance
c. Low labour input
d. Economic energy return

Results obtained from the digester systems outlined in this paper suggest that the fully mixed reactor has the ability to achieve these aims.

Several years experience operating these systems also demonstrates the following :

1) The ability to utilise high solids feedstocks of over 15% T.S. on a continuous basis.

2) Tolerance to variable loading rates (e.g. no loading at weekends and holidays).

3) Tolerance to a variety of feedstock mixtures.

The development of practical high solids digestion is conducive to the increased use of digesters in community and co-operative ventures, and integrated energy systems.

ACKNOWLEDEGEMENTS

A Grant for "Support for Innovation" from the Department of
Trade & Industry toward the installation costs of Digester
System C is gratefully acknowledged.

REFERENCES

Badger, D.M., Bogue, M.J. and Stewart, D.J.(1979). Biogas
production from crops and organic wastes. 1. Results of Batch
digestions. New Zealand Journal of Science Vol 22 11-20.

Finck J.D., Goma G., 1983. Anaerobic degradation of
ligno cellulosic materials : description of a new continuous
process. Energy from BIOMASS. 2nd GEC symposium. App Science
pubs., A. Strub, P. Chartier & G. Schleser Eds. 1033-1038.

Jewell W.J., 1978. Anaerobic Fermentation of Agricultural
residue. Potential for improvement in implementation.U.S.
Dept. Energy Report No. EY-76-S-02-2981-7, available through
NTIS Springfield VA. 427 pp.

Wujcik W.J., and Jewell W.J., 1979. Dry Anaerobic
Fermentation, paper presented at 2nd symposium on
Biotechnology in Energy production and conservation,
Gatlinburg. TN October 3-5.

Small-scale Power Generation with Bio-gas and Sewage Gas

A. A. M. Jenje

Nedalo b.v., Industrieweg 4, 1422 AJ Uithoorn, The
Netherlands

ABSTRACT

For biogas to be used as fuel for gas-engines, the most important
characteristics are calorific value, knock-resistance and hydrogen
sulphide content. The requirements for the efficient operation of a
gas-engine are described and experience with several types of engines in
the Netherlands are given. With sewage gas, as long as the hydrogen
sulphide content is less than 200 ppm, the gas-engine can operate
without any problems or precautions. Operating experience with a 150kW
combined heat and power unit operating on sewage gas is described and an
economic appraisal is made.

C5.1

KEYWORDS

Wood-gas, biogas, sewage gas, gas-engine operation.

BIOGASES

There is a large potential for the production of biogas from cow and pig
slurry and from sewage. The process provides a source of gas for
producing heat and electricity and also deals with the problem of slurry
disposal.
The average composition of three most common gases are shown in Table 1.

TABLE 1 Average Composition of Biogases and Sewage Gas

Gas	CH_4	CO_2	H_2S	H_2O	N_2	H_2	NH_3
			(vol %)				
biogas from cows	55–60	40–45	0.5	5.5	2.0	1.0	10^{-3}
biogas from pigs	65–70	30–35	1.0	5.5	2.0	1.0	10^{-3}
sewage gas	66–67	33–34	0.15	2.0	1.0	1.0	10^{-3}

For the gases to be used as fuel for engines, they have to fulfil a
number of requirements. The most important are calorific value,
knock-resistance and content of hydrogen sulphide (H_2S).

Calorific Value

The calorific value depends completely on the methane content, as shown
in Table 2.

TABLE 2 Calorific Values at Different Methane Content

Methane (vol %)	Net Calorific Value (MJ/Nm^3)
100	35.8
80	28.6
67	24.0
55	19.7

Air/Gas Mixture

In order to burn the gas completely, a minimal quantity of air is
necessary. If the quantity of air is exactly sufficient to burn the
quantity of fuel, the mixture is said to be stoichiometric.

In practice, the air (A)/fuel (F) mixture will never be stoichiometric
(st); to indicate the real (r) air/fuel mixture in relation to the
stoichiometrical value, the air proportion figure lambda is introduced
where lambda = $(A/F)^r /(A/F)^{st}$. In practice under a number of conditions,
the lambda is 1.2.

It is understandable that the calorific value of an air/fuel mixture
depends on the lambda figure. The calorific values of methane and
biogas with different methane contents at a lambda figure of 1.2 is
shown in Table 3.

TABLE 3 Calorific Value at Lambda 1.2

Methane (vol%)	Net Calorfic Value (MJ/Nm^3)
100	2.88
80	2.82
67	2.77
55	2.70

It is remarkable that, at varying gas compositions, the calorific value
at lambda 1.2 differs minimally.

For comparison, calorific values of gas and fuels, all at a lambda value
of 1.2, are shown in Table 4.

TABLE 4 Calorific Values of Gases and Fuels at Lambda Value of 1.2

FUEL	Net Calorific Value (MJ/Nm^3)
hydrogen	2.80
methane	2.88
propane	3.08
butane	3.11
natural gas	2.85
biogas	2.77
woodgas	2.30
petrol	3.22
diesel	3.11

Knock-resistance

If a gas is used as fuel for an engine, dangerous knocking can occur.
This is caused by the self-firing and explosive burning of the remaining
gas mixture in the cylinder by the high compression of the piston
movement. This high local pressure increase by the high burning speed
of the gases results in the typical knocking sound of the engine and
causes a too high mechanical and thermal load. It is obvious that this
manner of operation has to be avoided.

The knock-resistance value of gas is compared with clear methane in
Table 5. The performance is better as the 'methane figure' increases.

TABLE 5 Knock-resistance Value of Gases

Type of gas	Chemical Code	'Methane Figure'
hydrogen	H_2	0
butane	C_4H_{10}	10
propane	C_3H_8	33.5
coal gas	(see note 1)	33
natural gas	(see note 2)	77
methane	CH_4	100
biogas	(see note 3)	133
sewage gas	(see note 4)	133.8

NOTES

Specification (vol%)

	CH_4	C_2H_6	C_3H_8	C_4H_{10}	CO_2	N_2	CO	H_2
1	19				3.5	17	12.5	44
2	84	5.6	1.7	0.7	1.6	6.4		
3	60				40			
4	65				35			

Hydrogen Sulphide Content

The hydrogen sulphide (H_2S) content in gases is shown in Table 6.

TABLE 6 Hydrogen Sulphide Content of Gases

Type of Gas	CH_4	CO_2 (vol%)	H_2	H_2S
natural gas	82.5	1	–	–
coal gas	25	18	54	–
biogas (cows)	55–60	40–45		0.5
biogas (pigs)	65–70	30–35		1.0
sewage gas	66–67	33–34		0.2

To avoid engine problems, firstly there should be a hydrogen sulphide filter in the biogas supply line and secondly, the condensation of the exhaust gases should be avoided. After burning of the biogas, the hydrogen sulphide forms sulphuric acid with the condensation water. This acid causes corrosion in the sump and the exhaust system. Copper and copper alloy are especially affected. It is recommended that the exhaust gas temperature should be above 120° C.

Running the engine without stopping or the use of clean (natural) gas for about ten minutes before stopping and during starting will extend the engine life considerably.

        BIOGAS

Tests have been carried out in the Netherlands on a number of generator sets operating on biogas.

The most important conclusion is that even if the biogas has a good calorific value and the gas engines are running correctly, the hydrogen sulphide causes severe problems, especially if the unit is started and stopped frequently.

A good filter to bring the quality of the gas within acceptable limits seems to need so much care that the farmer is not able to fit it into his daily working schedule. The experience was obtained with several engines on various gases as shown below:

engine	gas from
Waukesha	cows
Opel	pigs
Fiat	mixture of cows and pigs
Ford	cows

The conclusion drawn from this experience is that oil change has to be done far more frequently than with engines operating on natural gas, even when special oil with a high Total Base Number (TBN) is used. Stopping should be avoided because of the aggressive hydrogen sulphide. But because of unreliable gas production, there were a great number of stops.

Generally, good results can be expected if a biogas generator produces sufficient gas to run a generator set and if condensation of hydrogen sulphide can be avoided.

The biogas supply line for a generator set should contain at least the following filters.

   (a) a filter for large particles and water in the form of a
       closed tank where the gas enters and leaves at the top.
   (b) a dust filter with 5 micron openings
   (c) a ferro-filter to filter the hydrogen sulphide

These filters have to be refreshed frequently because of the rapid changes in their state and the danger of explosion.

The gas consumption works out at between 0.6 and 1.0 cubic metres per kWh. The overall efficiency of the unit will be between 66% and 75%.

Economic Appraisal

The economic appraisal of units operating on biogas shows that the cost of the installation, including the biogas generator, gives a neutral situation in which the energy cost savings are equal to the running costs but there is no pay-back of the investment or of interest charges on the capital used. The main advantage is the slurry disposal.

SEWAGE GAS

There is considerable experience of gas engines operating on sewage gas, dating from the 1930's with slow speed (maximum 750 rev/min) dual-fuel diesel engines.

Present-day spark plug ignited engines operate at 1500 rev/min on sewage gas and,if sufficient gas is not available, automatically switch over to natural gas. With sewage works now operating with professionally controlled processes, generally, as long as the hydrogen sulphide content of the gas is less than 200 ppm, the gas engine can operate without any problems or precautions.

CHP Installation

An example of gas-engine operation on sewage gas is the 150kWe NUTEC
combined heat and power (CHP) installation at the Riool Watar Zuivering
Institut de Bilt in the Netherlands (Fig. 1)

Fig. 1 NUTEC CHP installation, De Bilt

As a result of the energy saving methods which are used, the electrical
energy demand is 150kW in the winter and 100kW in the summer. The power
output of the NUTEC unit is adapted to these values.

Normally, ferrochloride is added to the biomass in order to reduce the
hydrogen sulphide content from an average level of 2000 ppm to less than
150 ppm. About 40 litres is added to 3600 cubic metres of biomass each
week. There is a danger that an overdose would kill the bacteria. The
mixing of the biomass is done by gas-lances blowing gas under pressure
into the mass. Currently tests are being carried out with the sludge
from a water process, which contains iron, being added to the biomass to
see if it can be used to reduce the hydrogen sulphide level. It would
be an ideal solution for the sewage gas production and for the dumping
problem of the sludge.

For each cubic metre of gas, the CHP unit generates 1.8 to 2 kW of
electrical power depending on the load, and the heat is used for the
biomass. In summer, the exhaust gases are by-passed from the exhaust
gas heat exchanger because there is no need for the heat. The gas
supply line has several water locks and a fine gas filter.

The CHP unit is controlled each day from the metering. An inspection
for oil and water leakages is made every week and a minor maintenance
service is carried out every 1000 hours (spark plug replacement and
valve clearance).

The life of spark plugs is about half the life obtained with natural gas,
but there are high spark plugs that will last for at least 1000 hours.

The economic appraisal for this CHP unit is as follows:

Net investment	Dgld	275,000
Electrical output:		
summer 5600 x 0.18 x 100	Dgld	100,800
winter 2500 x 0.18 x 150	Dgld	67,500
Total	Dgld	168,300
Maintenance costs:		
8100 x 3.75	Dgld	30,375
Total saving	Dgld	137,925
Pay-back time 24 months		

Note: Dgld 3.75 per hour for maintenance costs includes overhaul and
      replacement of the engine at the end of its service life.

ACKNOWLEDGEMENT

The author would like to express his gratitude for information received
from Ir. R.R.J. ter Rele of I.W.T.N.O. (Internationaal Wetransport
Toegepast Natuurkundig Onderzoek) on gas-engine operation on biogas,
project No.700330107, Ing. P Hoeksma of IMAG (Institut Mechanisatie,
Arbeid en Gebouwen) on generating of biogas and using at farms,
publication No. 199,
Mr J Arkenhout and Mr P. Hoeksma of IMAG on manure fermentation, June
1984,
Mr F.P. Spreet of R.W.Z.I. de Bilt on the operation of their CHP
installation and M.W.M. for information on gas engines dated 1984.

Biofuels: the Domestic Market in the UK

R. Scott*, C. P. Wait**, M. P. Buckland**, G. J. Lawson* and T. V. Callaghan*

*Institute of Terrestrial Ecology, Grange-over-Sands,
Cumbria LA11 6JU, UK
**ADAS, Ministry of Agriculture, Fisheries and Food,
Government Buildings, Cop Lane, Penwortham,
Preston PR1 0SP, UK

ABSTRACT

Despite the renewed use of wood as fuel in the past 10 years, the UK market for wood and other biofuels is casual and poorly quantified. Over 200 000 woodstoves have been sold, to about 1% of households. Most owners fall back on solid fuel because of inadequate storage, quality and supply of wood. Straw and other organic wastes are used, mostly on farms and in factories. Briquetted fuels retail at over £130/t; good quality fuelwood is still less than £50/t. Changes in patterns of land use induced by reduced subsidies may stimulate the fuelwood market and persuade farmers to manage woodlands for fuel. A survey of woodstove owners in a rural area of Lancashire has shown potential demand, and a woodland management scheme is in progress on a local farm. Stove owners have contracted for wood at a fixed price, sufficient to meet management and production costs. Woodland quality should improve and new business be created in the community. Fuelwood catchments and markets probably exist close to many urban areas, but supply networks may be less economic where housing density is very low. Opportunity cropping may continue to be the pattern of extraction in the latter areas.

KEYWORDS

Biofuels ; fuelwood ; briquettes ; fuel costs ; fuel markets ; woodstoves ; land use ; woodland management ; coppicing.

INTRODUCTION

Biomass is a diffuse source with low efficiency of solar energy conversion. Bioenergy has had relatively minor funding for research and development in the UK, as it was thought not to apply to regions of denser population. However, Price and Mitchell (1985), Frazer (1982) and Ager (1984) have all observed that the commercial sale of biofuels is most viable in moderately populated areas. In very low housing densities the distance of transport makes centralized distribution less economic, so opportunity cropping of nearby wood resources (Mitchell, 1981) could satisfy demand. The

scale of use of the bioenergy resource will depend on the value of the fuels
to the consumer.   The decision to use a particular type of fuel is
related to several factors:
1. low cost and aesthetic aspects
2. availability and reliability of supply
3. convenience and safety
Briquetted fuel suppliers have achieved the second and third objectives but
have used mainly non-economic benefits to persuade potential buyers on the
first.   The three objectives must also be followed by the fuelwood trade,
but at the moment low cost and the sentimental interest in woodburning are
the main attractions. The production cost of fuelwood is above the present
market price, and wood packaged and sold to the standard of briquettes could
cost as much. More practical work is needed on on the economics of the whole
operation, and reported here are the early stages of a local investigation of
wood production and the fuelwood market for a farm woodland in England.

UK MARKET FOR DOMESTIC BIOFUELS

Frazer (1981) has estimated the fuelwood resource available in Scotland,
which has a relatively low population density. Better woodland management
and more productive trees would increase the number of households which could
be supported from 10 000 to 50 000.   A greater land take for fuelwood crops
as advocated by Stott and co-workers (1981), hastened by reduced agricultural
subsidies (Bowers and Cheshire, 1983) and new inducements for energy crops
(Seligman, 1985) would increase production and mean that many more homes
could in the future rely on wood as their main energy source.   Smoke
regulations and storage problems could rule out the use of wood in cities,
although it is technically feasible for biofuels to burn cleanly, and
briquettes can substitute for charcoal (Jones, 1984).

It is difficult to assess the actual level of biofuel consumption, or to give
it a precise economic value. The resources of vegetation necessary to fuel a
large number of households can be calculated. Taking 1% of UK land area
(2000 sq.km.), a notional yield of dry biomass of 5t/ha/y would give a total
of 1Mt/y. This amount would provide 5t of fuel for 200 000 households
(about 1% of the total) at an annual cost of £250 per house, assuming a fuel
price of £50/t.   This compares favourably with the cost of other forms of
heating, and as fuel costs are said (Frazer, 1981) to take up 6% of the
typical family budget, even very low income houses could benefit from wood
fuel. An industry with an annual turnover of £M50 would be valuable to the
rural economy.

Good quality logwood is easy to handle and safe to store.   Yet the price of
wood delivered in bulk is less than £50/t.   Its disadvantages are that a
year's supply is bulky and needs dry storage at the user's home. Also, in
this country, wood is usually obtained in an unfit condition for efficient
burning without further drying. Green wood (over 20% water content) burns
acceptably on open fires, where its main purpose is probably decorative. Its
heat value is abysmally low (Todd and Elliffe, 1983) and in any case open
fires can have negative efficiencies of up to 10% (Mueller Associates, 1981),
by removing more heated air from rooms than they supply - hence the interest
in enclosed stoves. Modern stoves are designed to a specified heat output.
Using wet wood is inefficient, corrodes boilers and blocks flues.

Briquettes of wood waste, straw and domestic refuse are available in the UK
but the trade has developed slowly.   Unfortunately, despite the 7Mt of straw
which is burnt off each year (Martindale, 1984) in the fields, straw

briquettes have met consumer resistance because of their high ash content. Briquetted straw/wood-sawdust mixtures are an acceptable product, and briquetted slabwood has been widely marketed. The retail price of the packaged fuels is over £130/t, the same as high grade solid fuels, but with 30-40% lower heat value. The price is justified by the cost of chopping, briquetting, packaging and distribution; feedstock is bought in at around £15/t. These fuels are a novelty and are sold through stove suppliers, fuel stations and garden centres. They compete with briquetted peat (Robertson and Godsman, 1984) which is usually cheaper, but has higher ash content. Refuse-derived fuel has found a minor place in the municipal sector. Briquetted fuels are consistent in quality, clean and easy to handle, and burn very well, but prove expensive in use compared with other solid fuels. A number of bale-burning furnaces (Campbell, 1984) are in use on farm in the UK. With the exception of charcoal for barbecues, fuels derived by conversion methods such as gasification, digestion and fermentation have not entered the domestic market in the UK in significant volume.

LOCAL MARKETS FOR BIOFUELS

Most bioenergy studies in the UK have focused on the management and productivity aspects of energy crops, not at local demand for fuel. Unwin and Mitchell (1982) assessed resources at farm level, and Brandon and Mitchell (1983) predicted that it could be profitable even in lowland farms to invest in energy forestry on suitable land. In India, a study of patterns of local energy consumption (Kumar, Vasudevan and Patwardhan, 1985) found the large extent of non-commercial energy consumption. Although the village social structure is different (the typical UK woodstove owner lives in a gentrified older house and is unlikely to be a manual worker), the conclusion may be much the same: that is only a small proportion of biofuels pass through the commercial sector. Supplies will be waste timber, fallen trees, wood yard offcuts and the wayside gleanings of the Volvo silviculturalist. Wood-gathering activity will absorb much leisure time, but unless the owner takes the trouble to ensure a good supply of well-seasoned timber before winter, or relies only to a minor extent on firewood, fuel will soon run out. Few owners in Britain have contact with a regular supplier of fuel and even fewer have an adequate space for storage.

Based on information from the woodstove trade, the British market is of the order of 50 000 units a year. There are about 40 listed WARM (Wood and Solid Fuel Association of Retailers and Manufacturers) dealers and probably an equal number of non-members selling stoves. The woodstove market has been active for about the last 8 years, and 200 000 units have probably been sold in that time. Doubtless some homes have several. However, it is inconceivable that all the stoves have supplies of wood to match their full burning capacity, as this would be around 1Mt/y. The extent of the present fuelwood market is estimated at 250 000t/y (Price and Mitchell, 1985). In fact the owners rely on other forms of heating or switch to solid fuel. Less than 5% of stoves sold are of the wood-only type, as owners prefer the reassurance of falling back on other fuels. Some even convert stoves to gas.

Mitchell (1983) has put potential production from residues from existing forests at 1.5Mt, so in theory all present demand could be catered for if the fuel could be produced at a price which would give an economic return on inputs. Land availability models have shown the potential for changes in land use and farmers will be looking at other ways to use their land. Types will include existing woodland and marginal and difficult parts of farms. Coppice woods are best-suited to lowland sites and single stem short rotation plantations of trees to uplands.

LANCASHIRE FARM WOODLAND PROJECT

A Lancashire woodland survey revealed a serious decline in the condition of
broadleaved woodlands.   Prompted by the findings a project was set up with
the following objectives:

    1. Improvement of management in order to retain woodlands
       of mixed age for their landscape value and wildlife habitats
    2. Provision of extra income for farmers by managing the woodland
       areas on their farms
    3. Identification and creation of a market for a regular supply
       of fuelwood and timber in the area
    4. Extra employment in the countryside

The project began in March 1984 and, in addition to finding a suitable site,
as much information as possible was gathered on wood as a fuel (Table 1). The
site chosen is a 16ha woodland on a dairy and sheep farm in central
Lancashire. The wood was typical of many in the region, with standing timber
of very poor quality, the only market for most of which being firewood.

TABLE 1   Wood as a fuel - assumptions used in the project

Storage space required	$1.75 \ m^3/t$
Maximum moisture content	20%
Seasoning time minimum	9 months
Size of log required	length 30-60cm; diameter 10cm
* Woodburning stove average consumption	4.7t air dried wood/y
Broadleaf coppice production average	3t/ha air dried wood/y
Land required to supply one woodstove	1.6ha coppice wood

* This assumes wood is not the sole fuel source

Much effort was spent in finding a long term market for low grade timber:
market demand for fuel was assessed in the area.   Enquiries started with
the local WARM member. He had sold about 3000 woodstoves in the previous 5
years. From the firm's invoices we mapped 52 stoves within a 16km radius of
the wood. All 52 owners were asked if they were interested in a regular
supply of firewood. Many of the stoves are at present burning solid fuel
because of difficulty in attaining regular and reliable supplies of ready
burnable wood. 26 replied and said they were interested. These people were
quoted prices of £14.50 and £24/m³ for wet and dry wood respectively. From
this group, 14 owners said they would be willing to form the woodburning
co-operative.   The advantage of the co-operative is that it provides a set
market for the farmer.   The woodstove owners benefit from long-term, ready
burnable supply of fuel without the problem of storage. Members can collect
the wood as they require it or have it delivered at extra cost.

The woodland management system chosen was coppice with standards, on a
coppice rotation of 10-15 years. This system was chosen because coppice
gives the best lowland fuelwood production (Pearce, 1980) and is relatively
simple to handle.   It can be tailored to the existing condition of the wood
and the preferences of the owner. A number of aims can be combined in the
management of the woodland.   The diverse structure with standards is

aesthetically attractive and good for wildlife. This project will provide working experience of the various facets of coppice production and assess the benefits to the wood. Felling and extraction are carried out by contractors. The cost of these operations might render the product unprofitable, but the aim is to establish realistic costs. Tenders were sent to 14 contractors for felling and extraction of 2ha of woodland. The average tender for felling, of the 6 replies, was £3000, and £4500 for extraction, splitting and stocking. We (ADAS) accepted a tender based on the volume of wood extracted, giving total costs of £1500 and £2500, at a rate of £4.50 and £5.50/m^3. Labour input can be very high in neglected woodland (Gascoigne, 1982) and only when regular coppicing is in progress can costs be expected to fall.

CONCLUSIONS

There is a shortage of good quality wood for woodstoves already in operation in the United Kingdom. Problems over storage and certainty of supply have led users to rely on solid fuel after their initial stock is exhausted. Early conclusions from trial coppicing plots show that the present market price for wood offers little short term incentive to the producer because it is artificially low. The establishment of reliable supplies of good timber will allow the price to find an economic level and encourage further investment in energy forestry, which should be increasingly attractive if farming subsidies decline. Taken together, waste arisings from farming, forestry and domestic sources would be able to supply a substantial part of the domestic solid fuel market. There are other less conventional sources such as natural vegetation (Callaghan, Lawson and Scott, 1981).

Fuel supply can increase to meet the installed capacity of stoves. In addition, wood and briquettes will be burnt on open fires as a luxury fuel. For convenience, the majority of the population will continue to use fossil fuels, principally natural gas, as their main domestic heating fuel. Those who at present use solid fuel should be aware of the range of renewable biofuels which could be used on their appliances. There would be many social benefits to the de-centralization of energy systems, not least the creation of work in rural areas. The active management of semi-natural vegetation for a variety of goals, including recreational, would be compatible with the present trends in the marginal areas of the UK. Pressures are developing in the EEC for a large proportion of the land producing food surpluses (about 9Mha) to be diverted to energy production. Whether the domestic biofuel market ever emerges from its present informal state remains to be seen, but energy cropping should be an aspect of planning for future land use.

ACKNOWLEDGMENTS

Thanks to Greg Garland, Tony Booth and Jonathan Brind for letting their brains to be picked on the subject of the woodstove market, and also to Simon Humphreys for information on wood as a fuel.

REFERENCES

Ager, B. (1984). Increasing market for firewood. In B. Andersson and S. Falk (Eds.)Forest Energy in Sweden. Swedish University of Agric. Sciences, Garpenberg. pp. 82-83.
Bowers, J. K. and Cheshire,P. (1983). Agriculture, the Countryside and Land Use. An Economic Critique. Methuen, London.

Brandon, O. and Mitchell, P. (1983). Comparitive Economics of Conventional and Energy Forestry in Great Britain. In A. Strub, P. Chartier and G. Schleser (Eds.) Energy from Biomass 2nd EC Conference. Applied Science Publishers, London. pp.132-136.

Callaghan, T. V., Lawson, G. J. and Scott, R. (1981). Bracken as an Energy Crop?. In D. O. Hall and J. Morton (Eds.) Solar World Forum Pergamon Press, Oxford. pp. 1239-1247.

Campbell, C. J. (1984). Experience in Marketing, Installing and Operating the Lin-Ka Automatic Straw Fired Boiler Systems. In J. Twidell, F. Riddoch and B. Grainger (Eds.) Energy for Rural and Island Communities III. Pergamon Press, Oxford. pp. 379-384.

Gascoigne, P. E. (1980). A Case for Coppice-with-Standards - for Profit and Pleasure. Q. Jl. For. 74 47-56.

Frazer, A. I. (1981). Wood as a Source of Energy for Rural and Island Communities. In J. Twidell (Ed.), Energy for Rural and Island Communities. Pergamon Press, Oxford. pp. 115-120.

Frazer, A. I. (1982). The Use of Wood for Industrial and Domestic Energy in Rural Areas of Britain and the Transfer of Wood-Energy Technology to Developing Countries. In J. Twidell (Ed.) Energy for Rural and Island Communities II. Pergamon Press, Oxford. pp.259-273.

Jones, M. B. (1984). Biofuel Development in Central Africa: Technology Transfer from Ireland. In J.Twidell, F. Riddoch and B.Grainger (Eds.) Energy for Rural and Island Communities III. Pergamon Press, Oxford. pp.355-361.

Kumar, T. S., Vasudevan, P. and Patwardhan, S.V. (1985). Pattern of Non-Commercial Energy Consumption and Availability in the Indian Domestic Sector - A Case Study. Agricultural Wastes 12, 55-60.

Martindale, L. P. (1984). The Potential for Straw as a Fuel in the UK. ETSU, UK Department of Energy, Harwell.

Mitchell, C. P. (1981). Wood Energy for Local Use. In J.Twidell (Ed.) Energy for Rural and Island Communities. Pergamon Press, Oxford. pp.99-103

Mitchell, C. P. (1983). The Potential of Forest Biomass as a Source of Energy in Britain and Europe. Int. J. Biometeor. 27 219-226.

Mueller Associates (1981). Wood Combustion: State-of-knowledge Survey of Environmental, Health and Safety Aspects. US Department of Energy, Washington D.C.

Pearce, M. L. (1980). Coppiced Trees as Energy Crops. Forestry Commission, Wrecclesham, Surrey.

Price, R., and Mitchell, C.P. (1985). Potential for Wood as a Fuel in the United Kingdom. ETSU, UK Department of Energy, Harwell.

Seligman, R. M. (1985). Biomass Fuels in a European Context. In W. Palz, J. Coombs and D.O. Hall Energy from Biomass 3rd EC Conference.

Stott, K. G., McElroy, G. H., Abernethy, W. and Hayes, D. P. (1981). Coppice Willow for Biomass in the UK. In W. Palz, P. Chartier and D. O. Hall (Eds.) Energy from Biomass 1st EC Conference. Applied Science Publishers, London.

Robertson, R. A. and Godsman, N. M. (1984). Peat as an Energy Source in Scotland. In J. Twidell, F. Riddoch and B.Grainger (Eds.) Energy for Rural and Island Communities III. Pergamon Press, Oxford. pp.395-402.

Todd, J.J. and Elliffe, M. D. (1983). Directory of Equipment for Industrial use of Crop and Forest Residue Fuels. University of Tasmania, Hobart.

Unwin, B. J. and Mitchell, C.P. (1982). A Study of the Feasibility of Coppice for Energy as a Lowland Farm Enterprise. In J.Twidell (Ed.) Energy for Rural and Island Communities II. Pergamon Press, Oxford. pp. 275-279.

TOPIC D

The Food and Agriculture Organisation (FAO) and United Nations Development Programme (UNDP) Projects

FAO/UNDP Integrated Energy Systems Project for Farms and Rural Areas

I. Hounam*, G. Riva** and J. W. Twidell*

*Energy Studies Unit, University of Strathclyde,
Glasgow G1 1XN, UK
**Institute of Agricultural Engineering, University of
Milan, Italy

ABSTRACT

Thirteen European countries, of both the western and eastern
blocs, are cooperating to develop renewable energy supplies for
farms. Each country should have a practical demonstration site
incorporating two or more types of supply. There is a common
methodology for technical and economic assessment. This paper
describes progress at the end of the first year. Already a wide
range of systems are operating, including biogas, solar heating
and drying, gas engines, heat pumps, aerogenerators, water pumps.

KEYWORDS   Agriculture, renewable energy supplies,
demonstration, economics

1   INTRODUCTION

1.1   General

The cost and availability of energy are limiting factors on
agricultural development. Across Europe there are wide ranges of
agricultural practice, environmental conditions, technical
support and financial resources. Nevertheless there is a strong
spirit for countries to cooperate and share experience. This
project is an attempt at such cooperation between 13 countries of
both the western and eastern blocs, with the aim of providing new
and economic energy supplies from renewable resources combined
with energy efficiency.

F1

The project is coordinated from the office for Europe, within the
Food and Agriculture Organisation in Rome.   The United Nation's
Development Programme offers financial support and a technical
directive from New York.   Each country has a small group of
representatives involved with their own demonstration site.
These national groups maintain contact with the corresponding
groups in the other countries by means of (i) correspondence,
(ii) visits from assessors, and, (iii) occasional European-wide
central meetings. This paper summarises the several independent
contributions made by delegates from the FAO/UNDP Project
countries to the Conference at Inverness.   It is not an official
document of the FAO or UNDP and has not been checked by the
delegates concerned due to shortness of time.

1.2 Conference Introduction

The aims of the projects are twofold:

1.   To determine economically viable technical solutions for
meeting energy demand in farms and rural communities from
renewable sources, and to identify combinations of such
technologies that give the best economic and operation results.

2.   To determine the barriers that hamper the diffusion of
renewable energy systems in agriculture and rural communities.

The FAO/UNDP project was started at the beginning of 1985, after
a preliminary two year study of common monitoring and assessment
methods.   The countries involved fall into two groups, Fig.1,

a) Seven less developed countries - Czechoslovakia, Greece,
Hungary, Poland, Portugal, Roumania and Yugoslavia.
b) Six more highly developed countries - Finland, France, Israel,
Italy, Sweden and the UK.   Of these only Finland, Italy and
Sweden had full participation in 1985.

The ultimate goal of the project is to permit the rural
communities of these countries to reduce their dependence on
external energy supplies, especially oil.   These supplies may be
transported long distances at great cost and usually entail
subsidies, implicit or explicit, because the rural area is
charged a similar price for its energy as in urban areas.   It is
intended to demonstrate that the use of locally available energy
resources reduces the production costs of agricultural
communities and so may increase farm profits.

1.2. Evaluating Economical and Technical Feasibility

For each participating country an analysis has been made of
individual technologies that are both technically and
economically capable of supplying energy requirements (excluding
transport) in agriculture and villages.   The methodology adopted
recognises the intrinsic limits - e.g. inflexibility in
operation or local availability of raw materials for each of the
available technologies.   The most rational combination for

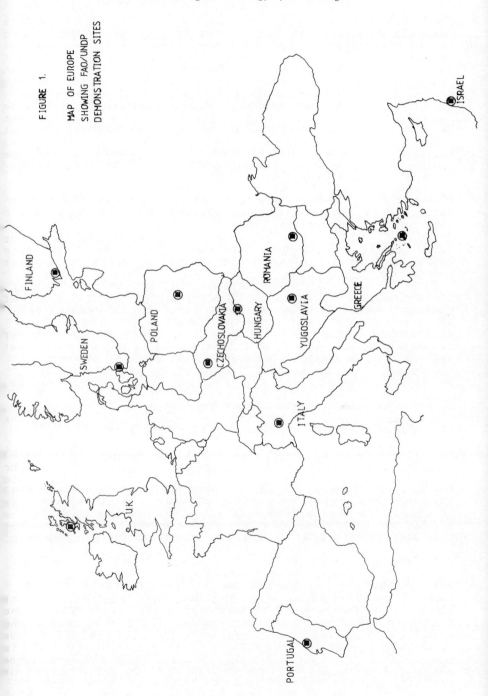

FIGURE 1.

MAP OF EUROPE
SHOWING FAO/UNDP
DEMONSTRATION SITES

reducing the external energy consumption, depending on the specific requirements and availabilities, is the goal of the analysis.   It is recognised that rarely will complete substitution for conventional fossil fuel or electricity be achieved.   In areas that have no centralised energy sources like grid electricity, the importance of developing local renewable supplies is clearly of even greater significance.

In order to realise these aims three steps were adopted:

1.   Demonstration sites, typical of more remote areas, were chosen in each country for the analysis and testing of one or more renewable energy technologies.

2.   A common method was defined to analyse the energy requirements of the sites and evaluate the locally relevant renewable energy supplies.

3.   A central computer model was made to simulate the renewable energy systems and calculate the optimum combination of plant sizes.

The computer model (SIENA) is used to evaluate all possible combinations of a range of energy generating plant and plant dimensions. SIENA takes into consideration the varying nature of the farm energy demand and matches this to the varying renewable energy supply. Any demand that cannot be covered is made up from conventional sources. Three possible strategies for the selection of the optimum Integrated Energy Systems (IES) are possible:

1.   Maximum contribution of renewable energy

2.   Maximum financial internal rate of return (IRR)

3.   Maximum contribution of renewable energy within set limits of IRR.

The current situation for the ten participating countries is that the first two steps have been completed. The sites have been selected, the energy demand curves have been measured and the locally available renewable energies have been identified. The next step is to define the optimum Integrated Energy Systems.

1.3 Barriers

Work on this part of the project has only recently started in earnest. A questionnaire has been designed to collect data on the positive and negative factors that affect the acceptance of renewable energy at the farm level.   These factors include legislation, subsidies, availability of equipment, and infrastructure for maintenance.

The reason for collecting these data is to define the areas of research necessary to overcome these barriers.   Also the results will identify how the governments of the participating countries may encourage the diffusion of renewable energy to rural areas.

2.1   POUCHOV COOPERATIVE FARM, CZECHOSLOVAKIA

2.1.1   Team Leader   -   H Jelinkova, Research Institute of
                            Agricultural Engineering, K Sancum 50
                            163 07 Prague 6, Repy.

2.1.2   Site

The Pouchov Cooperative Farm is an agro-industrial cooperative of
2350 ha situated 105 km east of Prague.   It is a typical farm
with many counterparts in the rest of Czechoslovakia.   Activities
include both agriculture and wood processing.   The farm
encompasses 7 villages with 472 active workers.   Wheat, barley,
rye, potatoes, sugar beet, fodder crops, fruit and vegetables are
grown on 1700 ha of agricultural land.   Animal production is
mainly for milk.   880 dairy cattle produce 3.8 million
litres/year of milk.   Flowers, vegetables and, in spring,
vegetable seedlings are grown in greenhouses of total area $4500m^2$.

2.1.3   Energy Consumption

The current use of energy for non-transport purposes is given in
Table 2.1.   This refers to only the part of the farm that is
being considered in this study.

TABLE 2.1   ENERGY DEMAND OF POUCHOV FARM, CZECHOSLOVAKIA

	Thermal	Electricity
	TJ/y	TJ/y
Forage Drying	0.63	0.09
Space Heating	9.32	-
Greenhouses	2.60	0.04
Workshops and Stables	-	0.79
Total	12.55	0.92

2.1.4   Installed Renewable Energy Systems

(a) Solar water heaters.   248 m^2 of solar collectors have been
installed for greenhouse soil and space heating.   The primary
circuit, filled with water and antifreeze to allow winter
operation, connects the collectors to a 44m^2 heat transfer area
counterflow heat exchanger.   Circulation is by a 1.5 kW electric
pump.   The secondary circuit consists of 42.6 m^3 storage tanks
connected to the greenhouse heating systems (a 4 kW electric pump
is used to circulate the water).

(b) Solar pond.   A 150 m^3 solar pond with surface area of 432 m^2
covered by plastic sheets has been constructed.   The temperature
range is + 25° to 2^{O}C

(c) Heat pumps.   Two heat pumps with a total rating of 30 kW (e)
and a Coefficient of Performance of 2.6-2.8 were installed.

F   5

FIGURE 2   LAYOUT OF THE INTEGRATED SOLAR COLLECTOR SYSTEM
POUCHOV COOPERATIVE FARM, CZECHOSLOVAKIA

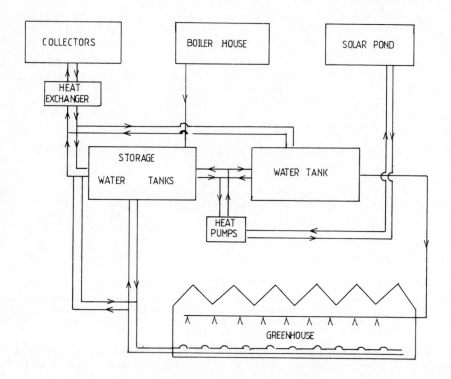

These can either utilise the heat from the solar pond or solar water heaters.   They are used to raise the temperature of the greenhouses heating water to above 45°C when solar radiation is insufficient on its own.

2.1.4 New technologies being considered.

A gasifier to fuel a generator set with a rating of 40 kW(e) is being considered. The fuel would be wood and crop residues. The engine cooling water would be used on site for greenhouse heating.

2.1.5. Integrated energy system

Fig. 2 illustrates the way in which the existing plant is interconnected. The system has only recently been completed and some testing was undertaken in a three month period in Summer 1985.

Solar collector efficiency with an ambient temperature range 15-30°C has been measured as 0.68 -0.60. The total efficiency of the system when warming irrigation water is in the range 0.40-0.43. In the period June to August as much as 70% of thermal energy needs were covered by the system.   The remainder of demand for heat can be covered by heat pumps.

2.2 GALINI AND EGARES VILLAGES, NAXOS ISLAND, GREECE

Team Leader   Prof Sp Kyritsis
                         Department of Agricultural Building
                         75 Iera Odos, Botanikos, Athens.

2.2.2   Site

Naxos is an island in the Kyklades group in the Aegean Sea. Agricultural land on the site is 300 ha, of which only 30 ha are adequately irrigated.   The total population at the site is 350 of which 50% is active.   The ambient temperature ranges from a yearly minimum of 0.5°C to a maximum of 37°C with an average of 18.5 $^{\circ}$C. An important development for the village would be to increase the proportion of the land under irrigation. Supply of renewable energy for pumping would be one means of achieving this. The   main crops are potatoes, vegetables, cereals and fruit.   Livestock production is mainly pig farming, with a pig population of 1900 animals. Also there are 600 goats and sheep, and 250 cattle.

2.2.3 Energy Consumption

Table 2.2 shows the present energy demand on the island of Naxos.

TABLE 2.2   ENERGY DEMAND IN GALINI AND EGARES, GREECE

	Thermal TJ/y	Electricity TJ/y
Greenhouses	0.58	0.06
Space heating	2.11	–
Process Hot Water	0.49	–
Water Pumping	1.97	0.13
Pig rearing	–	0.15
Houses	–	0.46
Total	5.15	0.80

The site is connected to the island's electricity grid. Currently electricity generation is fuelled by diesel and, to a lesser extent, LPG and wood.

2.2.3 Installed renewable energy systems

(a) Solar Collectors.   10 solar panels of $2m^2$ each have been installed for the supply of domestic hot water. A $500m^2$ greenhouse has been installed with a passive heat storage system. Heat from the upper level of greenhouse is stored in the subsoil for subsequent release at night (using a heat pump).

(b) Wind Turbine Generator.   A 10 kW(e) aerogenerator is currently under construction. The average wind speed at the site is 6.9 m/s, and so there is an excellent opportunity to generate power.

(c) Biogas.   An anaerobic digester for sewage disposal has been installed.   This is primarily for environmental health reasons. There are also benefits from energy saving by reclamation of clean irrigation water, and from the methane production.

Energy saving developments include:

a) Heat pump for greenhouse heating,
b) Plastic film for greenhouse cover,
c) Saving waste from irrigation water.

2.2.4 Other technologies being considered

The site is very suitable for windpower and investigations will be made into the best sites.

Hydro power from a stream situated above the village, at a height of 180 m with a flow rate of $100m^3$/h, shows promise as a renewable energy source.

An energy management system is being considered to control the whole system. Communication between the scattered user points and the central control system would be by 2-way UHF radio links.

2.2.5 Barriers

(a) Greenhouse heating. In Greece generally, this uses a large amount of the agriculture sector's primary energy consumption and is a prime target for energy saving. Geothermal heat could potentially be harnessed but little use is made of this source. The barriers seen as stopping adoption of geothermal power are -

a. The saline hot water requires expensive stainless steel pipework and equipment.

b. The most readily available sources provide a relatively low temperature.

The high cost of solar collectors and absence of low cost storage systems is seen as the main obstacle to widespread adoption of solar technology. The economics of solar heating could be improved by systems meeting part demand instead of providing a large overcapacity that is utilised only in rare circumstances. The same comments apply to domestic hot water systems. Despite the fact that only small temperature differences are needed for greenhouse heating, the heat pump is still too expensive to gain popularity.

2.3 RADOCZI COOPERATIVE, SZECSENY, HUNGARY

2.3.1 Team Leader -   Mr J Flieg
                         NationalInst of Agricultural Eng.
                         Tessedik Su 4
                         H 2101 Godollo

2.3.2 Site

This large farm cooperative of 7760 ha encompasses many villages, but only a small part of the farm including workshops and cattle raising has been considered in this project. Principal crops include cereals and forage on 40% of the total area. 4500 cattle are housed in sheds.

2.3.3 The current energy consumption is shown in Table 2.3 for the part of the farm considered.

TABLE 2.3   ENERGY CONSUMPTION   RAKOCZI FARM, HUNGARY

	Thermal TJ/y	Electricity TJ/y
Drying	0.31	–
Space Heating	3.78	–
Milking	0.36	–
Workshops	–	0.69
Biogas Plant	–	0.13
Processing	–	4.08
Office	–	0.20
Total	4.45	5.10

2.3.4 Currently installed technologies

A 1800m^3 anaerobic digester has been installed on the farm
close to the cowsheds. The substrate is cattle manure which is
piped to the digester from the cowsheds.   The daily effluent
from 700 cows (60m^3, dry matter 8-10%) provides about 1700m^3 of
biogas (55-60% CH_4). The plant operates at 32°C with a retention
time of 30 days.

The biogas is used in heating workshops and corn drying. Until
recently poultry houses were heated and in the near future
electricity generation will be fuelled by the plant. An added
advantage is that fermented sludge is spread on the fields as a
fertiliser and soil conditioner.

2.3.5 Other technologies being considered

In the near future a wood gasifier and combined heat and power
unit will be installed with a power of 42kW. In order to utilise
excess biogas a CHP unit of 110kW will also be installed. To
allow flexibility of operation of these two units either unit
will be able to operate on either the biogas or the wood gas.

2.3.6 Barriers

In general, for wood and crop residue burners, solar collectors
etc., the main problems causing low profitibility are high
capital cost and short effective operation time. In the case of
crop residues, extra manpower and machinery are necessary for
collection and processing. The new technologies have not been
demonstrated to be reliable enough, hence the farmers retain the
old heating systems as a standby. Hungarian farmers prefer to
plough crop residues into the soil to maintain fertility rather
than burn them.

2.4 SAN COSIMO FARM, S CRISTINA BISSONE, PAVIA, ITALY

2.4.1 Team leader -   Prof G Castelli
                            University of Milan
                            Institute of Agricultural Engineering
                            Via Celona 2
                            20133 Milano

2.4.2 Site

The site is a 108 ha dairy farm on the lower plain of the
Lombardy. It is a characteristic highly intensive farm with high
levels of capital investment. 18 people live on the farm of
which 8 are employed. Maize is grown on 43.5 ha for both grain
and silage. Winter cereals (wheat and barley) are grown on 26ha.
280 dairy cattle produce 945 t/y of milk.   100 head of beef
cattle give 292 t/head per year meat production.

2.4.3 Current Energy Consumption

TABLE 2.4   ENERGY CONSUMPTION ON SAN COSIMO FARM, ITALY

Housing	0.49	0.11
Stable	0.22	0.16
Maize drying	0.31	0.06
Silos & workshop	0.36	0.05
Total	1.38	0.38

2.4.4 Installed technologies

Although the farm has no wood supplies of its own, the
surrounding area supplies low cost biomass. This is used in a
gasifier which fuels a 28kW generator powered by an air cooled
motor with no heat recovery.

2.4.5 Technologies being considered

This site has been investigated using the SIENA model and the
results are summarised for each of the considered systems.

Solar driers

A two stage process for drying hay, using low cost solar
collectors incorporated in the structure of the drier, is being
considered. Costs can be kept low enough for economic operation
despite the short hay drying season (5 months of non continuous
use) on the farm.

Using the SIENA model the relationships between plant size and
solar energy utilised was found to be non-linear. At larger
plant sizes the extra returns in energy utilisation become
smaller. A $200m^2$ collector gives the highest Internal Rate of
return (IRR), but a $600m^2$ collector gives the highest Net Present
Value (NPV).

The Chosen Integrated Energy System (IES)

Of the three optimum sizes of the above plants -

    $600m^2$ solar hay drier
    20 kW(e) CHP plant
    110 kW(th) biomass burning boiler

are combined into an IES, the model predicts the following

Electric load covered	86%
Thermal " "	88%
IRR " "	12.5%

If only the first two technologies are used, the economics are
better, but the load covered is less.

Electric load covered   86%
Thermal   "   "   75%
IRR   "   "   13.8%

2.5   SHAAR HANEGER REGIONAL AGRO-INDUSTRIES, ISRAEL

2.5.1 Team   Leader -   Dr G Felsenstein
                                 Agricultural Research Organisation
                                 The Volcani Centre
                                 Inst. for Agricultural Engineering
                                 P O Box 6, Bet Dagan 50-250
                                 Israel

2.5.2   Site

The agro industry cooperative is run by the kibbutz of the
surrounding area up to 50 km in distance. Cotton is a major crop
and the total area under this crop is 12000 ha.

2.5.3   Current energy consumption

The main thermal energy consumption at Shaar Hanegev are given in
Table 2.5.

TABLE 2.5   THERMAL ENERGY DEMAND AT SHAAR HANEGEV, ISRAEL

	TJ/y
Poultry Processing	.13
Potato Chips Plant	.99
Laundry	.22
Total	1.34

2.5.4   Installed technologies

Solar Steam Generator

A plant with concentrating solar collectors designed to raise
steam at 26 atm was installed.   This prototype has been
abandoned after 2 years operation because of its failure to reach
the design operating pressure.   Experience gained on this
experimental plant has allowed the company involved to design a
new and successful plant that has been constructed in the USA.

Solar Drier

The solar collector forms the roof to the drier through which
warm air (up to 60°C) is circulatd by fans.   Static electricity
building up rapidly in dry weather, on the polycarbonate cover of
the collectors has been a problem (attracting small dust
particles).   The light transmittance of the cover dropped by 30%
in two years of operation.   An improved plastic cover has now
been installed.

Biomass Fuelled Boiler

40000 tonnes per year of cotton straw is available at the cooperative, but as a pilot demonstration plant a boiler consuming 10 t/ha was constructed. Experiments were carried out to find the most suitable system for harvesting and baling the straw. The most cost effective method is derooting with chisels and harvesting with a bailer to make a 1.5 diameter compacted bale. A special trailer has been designed to follow the bailer. Hence the trucks used for transport need not drive on the field, so reducing soil compaction.

The economic analysis suggests that in large scale operation bales will cost 30 US$/tonne of dry cotton stalks. A boiler that operates more than 2500 hours/year will be more economical than a conventional liquid fuel system.

The furnace consists of a single grate with one central burning chamber. Two main problems have been identified in the 500 hours it has been operated.

1.   Air pollution caused by emission of solid particles with the smoke.

2.   Clogging of the grate by ash.

Bag filters and cyclones are being installed in order to comply with environment protection laws. The grate and calandria are periodically rinsed out with water, without interrupting the boiler's normal operation. Occasionally the grate is scraped manually to clear the more stubborn ash deposits.

2.5.5   Barriers

The relatively low price of conventional oil and gas in recent years discourages the use of biomass or solar energy.

2.6.   SAN JOSE FARM, TORRES NOVAS, PORTUGAL

2.6.1   Team Leader   Mr T Morbey
                                  Estacas Nacional de Technologia
                                     dos Prodotes Agrarios - INIA
                                  Quinta do Marques
                                  2780 Oieras

2.6.2 Site

This small farm occupies an undulating 48 ha site. Tree crops of figs and olives are grown on 30 ha. There are 4 ha of vineyards and the rest of the farm is devoted to grassland where a few sheep are grazed.

2.6.3 Current Energy Consumption

The farm's energy consumption is shown in Table 2.6

136   I. Hounam, G. Riva and J. W. Twidell

TABLE 2.6   ENERGY CONSUMPTION ON SAN JOSE FARM, PORTUGAL

	Thermal	Electricity
Housing	0.22	18
Workshop	-	4
Distillery	1.70	4
Water Pumping	-	7
Total	1.92	33

2.6.4 Installed Technologies

A 200m^2 solar hot air collector with 440m^2 of drier area has been installed. The collector section is built on a south facing slope. It should be noted that there is no clear demarcation between the collector and the drier area which is covered with the same plastic sheeting, but is not at the optimum angle. The exit chimney part of the structure is constructed of black corrugated metal to act as a further solar collector. This causes convection to aid the air circulation through the drier which is controlled by a damper situated on top of the chimney. The drying capacity is 1.2t of fresh grapes which yield 0.25-0.3 raisins.

The raisins produced in this way are of such a high quality compared with those produced by the conventional method of open-air drying, that they fetch a much higher price on the market. The extra profit in the first year of operation was sufficient to recover all the capital and installation costs. Out of the drying season the structure can be used profitably as a greenhou

2.6.6 Other technologies under consideration

Wind turbine water pump.   The international cooperation in this project is illustrated by the fact that the Hungarian Government will supply a wind turbine pump to the farm.   This will be used to increase the irrigated area, particularly vineyards.

Biomass Residues Burner

The farm has a small distillery where spirit is distilled from wine and other fermented fruit.   It is currently the largest fuel consumer on the farm. Switching this sector to use biomass residues instead of wood would be of great benefit.

2.6.7 Barriers

With particular reference to solar driers, the results in Portugal have been varied in success.   The demonstration site where the coordination team have been involved from the planning stage, has been of great economic benefit to the farm. In another case nearby the farmer has not followed the design supplied and has constructed a structure that functions well as greenhouse but not as a drier.   This farmer has learned his lesson and is now asking for help to construct a better drier.

F

The conclusions drawn by Portugal's team leader on promoting solar driers are:

1. The correct drier for the site must be installed.

2. The scientists must explain the principles of the drier.

3. The farmer should be given an idea of the increased profits possible.

4. The government advisers to agriculture should thoroughly understand the techniques.

5. The farmers must be skilled in raisin production in order to select the correct species and store and pack the end product correctly.

2.7   SKONE, SWEDEN

2.7.1   Team Leader   Dr Ingvar Jansson
                                    Swedish University of Agricultural
                                    Sciences
                                    Dept of Farm Buildings
                                    P O Box 624, 022006 Lund.

2.7.2   Site

The farm has 70 ha of arable land and is situated near Skone in the south of the country.   The main crops are grass, grain, oil plants, sugar beet and fodder peas.   The grain, barley and wheat, requires drying before storage.   The animal stock is 60 milking cattle which are kept in stalls in the winter and grazed in summer.

2.7.3   Current energy consumption

The total electricity consumption at the farm in the year 1982/83 was 50 MWh.   Power curves are available for individual processes on the farm, but no information on thermal energy requirements has been collected.

2.7.4   Currently installed technologies.

Heat Pump

An electric powered heat pump of 4.5 kW(e) has been installed to provide space heating and domestic hot water for a 250m^2 apartment with four inhabitants.   The C.O.P of the plant is 2.8-3.0 giving more than 12 kW of heat.   The payback time on the investment of 70000 Kroner is estimated to be five years with a discount rate of 20%.

2.7.5 Other technologies under consideration

Anaerobic fermentation is possible in the winter months when
cattle dung can be collected from the cowsheds. About
tonne/ha of straw could be collected for burning.

2.8   OKSANEN FARM, SOUTH FINLAND

2.8.1   Team Leader -   Dr E H Oksanen
                               Work Efficiency Association
                               Box 28, 00211 Helsinki.

This project on this farm, the name of which is coincidentally
the same as that of the country's team leader, is reported
separately in Dr Oksanen's paper, Conference paper CI.

3   Other sites and County Team Leaders

3.1   POLAND   Prof Tyminsky
                         Institute for Building, Mechanisation and
                         Electrification in Agriculture
                         Rakoviecka 32
                         02 532 Warsaw.

Two farms are being investigated by the Polish Team

1. Olejuk Farm, Izdebna.   This farm's main crop is maize
grown on 30 ha, and the principle animals raised are pigs (600)
A 16.5m^3 biogas digester for heat used in pig food preparation
and a 10m^3 biogas plant for domestic cooking have been installed
In the future a 20 kW aerogenerator could be constructed.

2. Lesniak Farm, Prazkow.   This has 15.5 ha of arable land
devoted to crops, trees and forage. 24 cattle are kept. A 20 k
aerogenerator supplying electricity for heating and power has
been installed. A 0.3. kW(e) heat pump is used for cooling
milk and water heating. Other systems under consideration are
biogas plant fuelling a CHP system.

3.2   ROMANIA   Dr D Teaci
                         Academy for Agricultural and Forestry
                         Sciences,
                         Boulevard Marasti 61, Bucharest.

The site is the "Enterprise 30 December", cooperative farm. Land
area is 3200 ha, 50% of which is devoted to crops, 40% fo
forage, 10% trees and 38 ha to greenhouses. Cattle pigs an
poultry are kept in very large numbers. Two 400m^3 bioga
digesters with CHP plants (30 kW(e) and 45 kW(e) have bee
installed. Further biogas plants are planned.

3.3   YUGOSLAVIA   Dr M Todorovic
                           University of Belgrade
                           Institute of Agricultural Engineering
                           Nemanjina 6, 11080 Zemun

The Yugoslavian site is Leservic Farm, which has a total area of
arable land of 8.5 ha of which 40% is given over to crops, 35% to
forage, 17% to trees and 6% to fuel wood. Cattle, pigs, and
poultry are kept. A 30m^3 biogas plant has been constructed and
supplies gas for lighting and cooking.   Solar collectors of 6m^2
area are used to heat the biogas digester. A CHP system is
planned to supply heat and electricity, 3 kW(e) to cover all the
farm energy demand.

3.4   UNITED KINGDOM

Energy use on the island of Eigg, West of Scotland, has been
studied. The community of 64 people now use a large variety of
imported fuels, and there is no grid electricity.   The detailed
results have been published (J W Twidell and A A Pinney, 'Energy
supply and use on the small Scottish island of Eigg', Energy,
vol.10, 963-975, 1985).   It is hoped to encourage the
development of significant renewable energy supplies of hydro,
wind, biofuel and solar power.

3.5   FRANCE

Progress is being made to establish a meaningful demonstration
site within the aims of the overall project.

4   CONCLUSION

This is the first general report published concerning the
FAO/UNDP project in Europe. Although several years of
development and analysis remain, it is already clear that some
renewable energy technologies are being accepted as economic (eg.
solar drying, biogas) despite a low general level of technical
appreciation.   The barriers to continued development vary
greatly between the 13 countries concerned, and are frequently a
result of financial policy rather than technical achievement.

As authors, we express our gratitude to the persons named in the
paper, and also to the Coordination Committee members, M Bernard
Biet (FAO, Rome), Prof. G Castelli and Prof. G Pellizzi
(University of Milan).

TOPIC E

Hydro and Tidal Power

Integration of Renewable and Fossil Fuel Supplies for Grenada, West Indies; Especially Considering Hydro-Electricity

P. N. Hooper

James Williamson & Partners, Consulting Engineers,
231 St. Vincent Street, Glasgow G2 5QZ, UK

ABSTRACT

To reduce dependence on imported fossil fuel, several renewable energy
options are under consideration. The first stage of a hydro power feasibility study is described and features of its intended integration with
existing and future fossil fuel systems.

KEYWORDS

Integration of renewable and diesel supplies; hydro-electric feasibility study;
Grenada, West Indies.

INTRODUCTION

Grenada is the most southerly island in the chain forming the Windward
Island Group in the East Caribbean. It is about the same size and shape as
Arran, in the Firth of Clyde, but has a population of 100,000 people, and
some 65% of households receive an electricity supply from the 11kV grid,
Fig. 1. The grid is fed from a single diesel fueled power plant of about
10 MW nominal capacity.

The power plant has a total of 9 diesel units installed, and 2 mobile units
on hire. The installation dates range from 1962 to 1984. Several of the
older units require major repair or are unreliable. The present firm capacity is 5.7 MW, and the maximum demand is 5.0 MW.

In recent years there has been significant load shedding, and several hotels,
restaurants and commercial premises have installed their own stand-by generating plant. By 1984 the average fuel cost per kWh had reached 6.5 pence,
straining the country's foreign exchange resources, and encouraging the
search for alternative energy sources.

E1.1

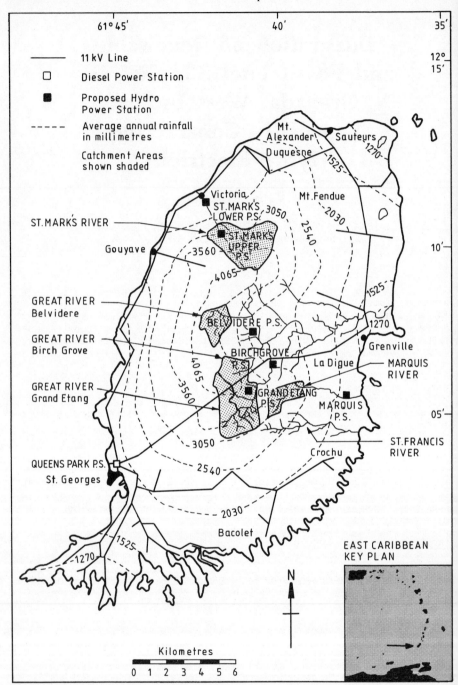

Fig. 1 Grenada–11kV Grid & Proposed Hydro Power Developments

E1.2

ALTERNATIVE ENERGY SOURCES CONSIDERED

In common with other East Caribbean islands, Grenada lies in the path of the
rainbearing north-east trade winds, and geologically is volcanic in origin.
These features, along with the island's mountainous topography, provide
potential for development of wind power, hydro electric generation and geothermal energy.

The Caribbean Meteorological Institute in Barbados has been collecting wind
data covering all the islands in the group. Pilot wind generator projects
are underway in Antigua and Barbados, and from studies which will monitor
these projects, it is hoped eventually to evaluate their commercial and
operational viability for use in other parts of the Caribbean.

A geothermal reconnaissance study was carried out in Grenada in 1981. The
data obtained seem to have been limited to surface exposures of the volcanic
formations, faults, and lines of fracture. This would identify potential
deep drilling sites if a search for hot rocks is to be carried out. At the
present time the technology and economics of power generation from hot rocks
is considered too innovative to include in a medium term generation programme
appropriate to Grenada's needs.

A pre-feasibility study of Grenada's hydro-electric potential, by SCET
International (France) in 1981, identified 9 river catchments that might
support a total installation of about 7 MW. A similar study by Polytechna
Hydroprojekt, Czechoslovakia, in 1982 covered the same ground but reached
more conservative conclusions.

After considering the above options, Grenada Electricity Services Ltd., who
are responsible for generation and electrical supply distribution on the
island, decided that hydro-electric power offered the most certain prospects.

REQUEST FOR HYDROPOWER AND GENERATION DEVELOPMENT STUDY

In August 1983, Consulting Engineers were invited to submit proposals for a
feasibility study to be carried out in 3 phases comprising Hydro-Electric
Preliminary Study, Hydro Power and Medium Term Generation Programme, Detailed
Design of Identified sites. In October 1983 the Consultants, Kennedy &
Donkin, in association with my own firm, James Williamson & Partners, submitted their proposals for this work. However, on the 19th of that month
political trouble reached a crisis, and this was followed by military inter-
vention by the Organisation of East Caribbean States with United States
assistance.

As a result of the disruption to life on the island it was decided to implement
the first phase only and delay subsequent phases until future demand requirements could be estimated with more certainty. Work on the first phase, the
Preliminary Hydro-Electric Feasibility Study, was undertaken between May and
August 1984.

HYDRO-ELECTRIC PRELIMINARY FEASIBILITY STUDY

It was required to examine and report on the 4 most favourable hydro-electric
developments already identified from pre-feasibility studies, namely Great
River, St. Francis, Marquis and St. Mark's rivers, Fig. 1.

A desk study, followed by a field visit, evaluated hydrological and topo-
graphical data, assessed schemes, layouts, access and environmental considerations, surface geology and construction methods, from which an evaluation
of outputs and estimated budget costs were made. The report ranked the
schemes in order of merit and included recommendations for ongoing hydrological data recording, and site investigation work.

Figure 1 shows the average annual rainfall distribution for Grenada.
Prevailing wind is approximately along the major axis of the island hence
there is no rain-shadow as such, and rainfall distribution corresponds roughly
with altitude, the latter reaching to over 800m. Duration of the wet season
also varies directly with altitude, being almost unbroken on the central
ridge but only about 6 months in coastal areas. The volcanic origins of the
ground formation results in certain rivers having higher sustained flows
than others. This applies particularly to parts of Great River where the
Grand Etang branch is fed from the crater lake of the same name, and to
St. Mark's River, where inflow occurs from springs.

All the rivers are flashy, and each development would be of the run-of-river
type, comprising a small dam with an intake structure, similar to Fig. 2,
and a pipeline leading to a power station. The dam illustrated was being
constructed on the St. Francis River for a water supply. It is 4 metres high,
of a reinforced concrete cantilever design in general use in Grenada. It
has been built by labour intensive methods without the benefit of an access
road.

Fig. 2. Water Supply Dam - St. Francis River

For the proposed hydro development, the intake headponds would be sized to provide daily storage, dams would range in height from 3 to 5 metres, and the storage volume would allow about 6 hours generation during periods of low flow. It is visualised that dam construction and pipelaying would be carried out by direct labour, making use of expertise in this field already developed by the Grenada Water Commission.

Suitable construction equipment is available from Government and private sources. There are a number of local Contractors with sufficient capability to undertake power station construction. Access roads are expensive to construct and to maintain in the difficult terrain with high rainfall. Footpath access to intakes is visualised with power station sites located close to existing roads wherever possible. For the power station plant, it is expected to include supply, erection and commissioning in one contract with an overseas manufacturer.

In view of the major water supply abstraction works already under construction on the St. Francis River, the proposed hydro power scheme was not investigated further. The remaining schemes were all examined on the ground and power estimates prepared. Leading particulars are listed in Table 1, the total estimated budget costs being recovered over 30 years to arrive at an annual cost, operating and maintenance allowances are added to give an indicated unit cost of energy. The indicated cost of hydro generated units, is in the range 3.7 to 5.6 pence/kWh. This compares favourably with fuel-only cost for diesel generated units of 6.5 pence.

TABLE 1   Scheme Parameters & Budget Cost Estimates

DEVELOPMENT	Installed Capacity kW	Gross head m	Net Average Flow m3/s	Design Flow m3/s	Annual Energy Production MWh	Budget Cost of Project (1984) £ x 1000	£/kW	Estimated Energy Cost p/kWh	Turbine Type
GREAT RIVER Grand Etang	410	100	0.46	0.55	2,260	740	1,800	3.7	Francis
GREAT RIVER Birch Grove	670	91	0.87	1.05	3,700	1,400	2,100	4.3	Francis
GREAT RIVER Belvidere	215	92	0.28	0.34	1,220	570	2,600	5.3	Francis
MARQUIS River	295	275	0.13	0.15	1,630	530	1,800	3.7	Pelton
ST. MARK'S RIVER Upper	265	98	0.32	0.38	1,420	580	2,200	4.6	Francis
ST. MARK'S RIVER Lower	250	67	0.45	0.55	1,490	730	2,900	5.6	Crossflow
TOTALS	2,105				11,720				

It has been recommended that the 2 Great River developments, Grand Etang and Birch Grove, which would allow cascade generation, and the Marquis River development, with a combined installation of some 1,375 kW, should be included in the future power development programme for Grenada.

OPERATION AND INTEGRATION

It is intended that the second stage of the study will produce a generation development programme to meet the projected medium term demand requirements, using least cost solutions. Since this stage of the work has still to be implemented, the description given of operation and integration are general intentions, which would require to be optimised in detail.

Lack of storage for firm power, and the likely limit of available energy from hydro generation, determines that thermal back-up will be an ongoing requirement, operating in parallel with the hydro schemes. It is expected that the primary use of hydro power would be to replace diesel fuel which presently accounts for over 80% of generation costs. This would be achieved by giving priority to hydro generation whenever sufficient river flow is available, and particularly when flow would otherwise spill at intakes. Additionally, by providing some storage volume in headponds, the full hydro kW capacity can be brought in to meet peak demand even when river flows are low. A typical daily load curve projected for 6 MW peak, is shown in Fig. 3. It is seen that the proportion of base load would be quite low and demand would only exceed 4.5 MW for about 5 hours. The excess could conveniently be matched by hydro output. A preliminary flow duration curve, Fig. 4, based on experience elsewhere in Caribbean and Pacific tropical islands, indicates dry weather flow (95% exceedence) as about 30% of average flow, hence if a design flow of 1.2 times average is adopted, full power generation could be available daily for 6 hours if the storage volume can be provided. The flow duration curve indicates that theoretically 75% of river flow can be utilised for power generation on the above basis.

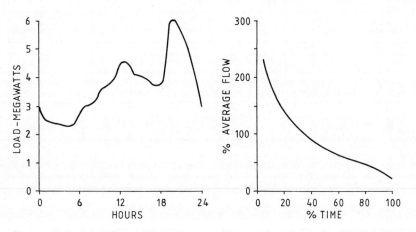

Fig. 3.   Daily Load Curve.(Typical)   Fig. 4.   Preliminary Flow Duration Curve.

At this stage, for budget cost purposes, it has been assumed that the hydro plants would have synchronous generators with automatic voltage regulators. It is possible that the second stage of the study may show that induction generators would be preferred, in which case costs would be reduced.

Most of the hydro schemes will have fairly long pipelines (Grand Etang 450 m, Marquis river 2,500 m), and it is hoped to minimise costs by dispensing with surge chambers and flywheels. Consequently the hydro schemes in the stand-alone mode, would not be suitable for frequency control, and it is expected that this function will be performed by the diesel units.

It is anticipated that the hydro plants will be operated by remote control from one power station for start up and shut-down, with provision for local control. A likely location for the control centre is Birch Grove, which would probably have the largest installation, and is well situated in relation to the existing 11kV grid and communications throughout the island.

The 3 developments selected from the preliminary feasibility study for further assessment have an average annual energy output potential of 7,600 MWh. The average annual gross generation, from diesel units over the last 6 years has been 24,500 MWh, hence nearly one third of diesel fuel costs could be saved by these 3 hydro schemes. This represents, at 1984 prices, £500,000 of foreign exchange that could be saved each year by hydropower, for which the investment cost recovery time might be only 5 or 6 years.

ACKNOWLEDGEMENT

The author wishes to thank the following organisations for information used in the preparation of this paper: Grenada Electricity Services Ltd., The Central Water Commission and The Land and Water Resource Unit, Grenada; The Caribbean Meteorological Institute, The North of Scotland Hydro-Electric Board, Kennedy & Donkin Consulting Engineers.

Tidal Energy Potential around the Coast of Scotland Using the Salford Transverse Oscillator

C. G. Carnie* I. D. Jones* I. Hounam*, G. Riva** and Twidell*

*Crouch & Hogg Consulting Engineers, 18 Woodside
Crescent, Charing Cross, Glasgow, UK
**Salford Civil Engineering Limited, Telford Building,
Meadow Road, Salford, UK

ABSTRACT

This paper outlines the development of a new type of positive displacement
ultra low-head hydropower machine called the Salford Transverse Oscillator
(STO). The STO is designed to operate at heads in the range 0.5 - 3m, and
has potential for development at run-of-river and tidal sites with installed
capacities in the range 50kW to 10MW.

STO operating characteristics have been used in a desk study of potential
tidal energy sites around the Scottish coast. Of 127 sites originally
investigated, some 49 were considered to have potential for development.
These sites range from small lochs with installed capacity of 250kW and
annual output of 300MWh to larger lochs with capacity of 7.4MW and annual
output of 11,000MWh. The total annual energy production from these sites
amounts to 142GWh from an installed capacity of 99MW.

This study had led to the adoption of a site at Sponish, North Uist in the
Outer Hebrides for prototype development of the STO. Here a 3 cell STO
barrage with an installed capacity of 270kW will be built for supply of
electricity to Lochmaddy Hospital in parallel with the North of Scotland
Hydro-Electric Board System. Construction at the site will begin in 1985
with commissioning in mid 1986.

KEYWORDS

Ultra low head hydropower; tidal energy potential; prototype development.

INTRODUCTION

The Salford Transverse Oscillator (STO) is a new type of positive displacement ultra low-head hydro-electric/hydro mechanical machine being commercially developed by AUR Hydropower Limited. It is specifically designed to
operate at tidal and run-of-river sites with hydraulic heads in the range
0.5 - 3m, which are generally considered to be too small for economic development using conventional hydro-electric equipment. At such

low heads, large water flows are required to produce worthwhile amounts of
power which makes conventional rotodynamic machinery expensive and possibly
ineffective. The STO, however, by virtue of its simple design and large
water passages, provides a solution to this fundamental problem.

The research and development work for the STO has been carried out at
Salford University.

DESCRIPTION

The STO was first conceived by Professor E M Wilson and Dr G N Bullock of
Salford University during development work for AUR Hydropower Ltd on another
type of ultra low-head hydro-electric generator.

The main features of the STO are shown in Fig 1. It consists of a parallel
double wall barrage built across the direction of flow in a river, canal or
tidal estuary, to make use of or create a hydraulic differential. At regular
intervals along the barrage are gaps into which are placed gates, the
sections of barrage between the gates are called cells. The purpose of the
gates is to direct water through the cells from the higher level to the
lower.

Paddles are placed in the cells to prevent the free passage of water from
upstream to downstream. There is, thus, a net hydraulic force on a paddle,
in the direction the water moves, causing it to be pushed to the end of the
cell at which point the gates are triggered reversing the flow and paddle
direction. By linking the paddles at the top to a carriage and synchron-
izing gate action, it is possible to make the carriage and paddles oscillate
from end to end of the cells along the barrage. It is this motion trans-
versely to the main direction of water flow which gives the STO its name.

A power take-off rod links the paddles at about one-third of their height
above the base of the cell. It is located to pass through the centre of
pressure of the water acting on the paddles and so keep the bending moments
obtained at the carriage to a minimum. The rod, which passes through the
centre of the gates but does not impede their motion, is linked to a double
acting hydraulic ram. The ram provides resistance to the motion of the
paddles causing them to work by pumping pressurized hydraulic oil.

The STO gates are driven by actuating rams which are supplied by a small
proportion of the high pressure oil 'bled-off' from the main hydraulic
circuit, making the STO self-powered.

In most cases the pressurized hydraulic oil would be passed via accumulators
which smooth the flow, to variable displacement hydraulic motors linked to
electric generators. However, the oil can be used directly for driving
hydraulic machinery, or to produce heat by being (blown) through a restric-
ting valve and heat exchange unit.

MODEL TESTS

The period of operation, discharge, power output and efficiency of the STO
are dependent upon its physical dimensions, the hydraulic head, the upstream
and downstream water levels and the force or loading applied to resist the
paddle motion at the power take-off rod. Extensive tests have been carried
out at Salford University to determine the relationship between these parameters and to dimension a unit which is most likely to produce energy at

1 — CARRIAGE   WHEELS
2 — PADDLE
3 — WALKWAY
4 — GATE
5 — POWER TAKE OFF CYLINDER
6 — BARRAGE WALLS

7 — CARRIAGE   FRAME
8 — MACHINERY   HOUSE

UP—UPSTREAM   WATER   LEVEL
DOWN—DOWNSTREAM   WATER   LEVEL

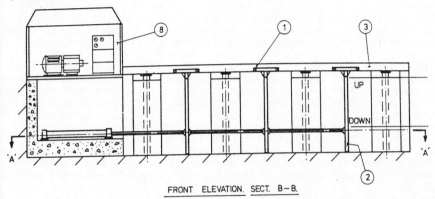

FRONT ELEVATION. SECT. B—B.

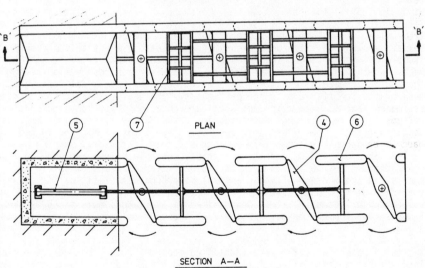

PLAN

SECTION A—A

SALFORD   TRANSVERSE   OSCILLATOR

A.U.R. HYDROPOWER LIMITED.   1984   FIG. 1

E2.3

154   C. G. Carnie *et al.*

minimum cost. The model parameters showing the best results are given
below:-

>   3 cells
>   cell length   = 0.88m   cell depth = 0.88m
>   cell width   = 0.44m   gate width = 0.22m
>   stroke length = 0.78m

A diagrammatic representation of a set of results for a constant downstream
water level is shown in Fig 2. It can be seen that it is necessary to place
a specific load on the paddles to ensure optimal power output. The discharge
through the STO is dependent upon the prevalent water levels, the period of
the carriage oscillation and the speed of gate operation. Accordingly,
since for any given system the sizes of the power take-off and gate actuating rams are fixed, operating pressure determines the applied load and so
discharge is specified at any point within the diagram. A series of these
diagrams can be produced for different values of downstream water level and
thus the operating parameters for the STO under any condition may be
specified.

Further, a STO operating characteristic to give optimal power under any condition can be computed by producing a family of optimal power envelopes for
various downstream levels, noting variation in applied load and discharge
along each curve, and then superimposing each curve onto a single diagram
Fig 3.

    DESK STUDIES

An examination of the tidal range around, and geography of, the Scottish
coastline suggests that there are numerous locations where a tidal power
plant might be operated: the sea lochs are often characterized by narrow
necks and shallow entrance sills which would make the construction of an
impounding barrage a relatively simple matter. The mean tidal range around
much of the Scottish coast, and particularly the west coast and the Hebrides
is in the range 3 - 4m with smaller tidal ranges in the Orkney and Shetland
Islands and a larger range exceeding 7m in the Solway Firth. This implies
that, in order to allow economic development of sites, a machine is required
which will produce worthwhile power outputs from continually varying heads
in the range 0.5 - 3.0m and is, preferably, capable of two-way ebb and flood
generation. The STO offers a, hitherto unavailable, means of developing
such sites and consequently the optimal power characteristic of Fig. 3 was
scaled-up using Froude Laws (with zero majoration) for use in energy analysis
in a study of potential tidal power sites in Scotland commissioned by the
Scottish Development Agency and the Highlands and Islands Development
Board.[1]

Initially potential sites were identified from the O.S. 1:50,000 scale maps.
The suitability of individual sites was then decided by reference to Admir-
alty Charts and Tide Tables, having regard to required barrage length,
enclosed basin area, depth at barrage and the size and incidence of tidal
ranges. Data from suitable sites were then used with STO operating characteristics in computer programs developed at Salford University to optimize
the number of STO units installed and compute annual energy production for
double effect operation.

Of 127 sites examined in the study, some 49 sites were considered to have
potential for development using a standard size of STO unit of the
following dimensions:-

E2.4

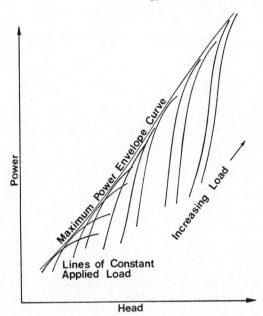

STO Power Output Against Head for Constant Downstream
Cell Depth and Varying Applied Load

FIG. 2

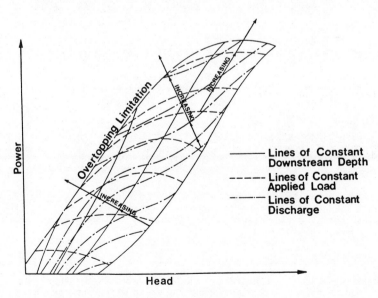

STO Operating Characteristics   FIG. 3

156   C. G. Carnie *et al.*

```
3 cells per STO unit
cell depth        = 6m      cell length   = 6m
cell width        = 3m      gate width    = 1.5m
stroke length     = 5.3m
```

The locations and details of these sites are given in Table 1;
they range from small lochs for which the installed capacity might be as low
as 250kW and the annual energy production 300MWh, to larger lochs with
installed capacity of up to 7.4MW and the annual energy output of 11,000MWh.
The total annual energy production from these sites amounts to 142GWh from
an installed capacity of 99MW.

Two sites, Loch Houram in North Uist (Sponish), Site No 102, and Loch Ceann
Hulavig (Callanish), Site No 126, were surveyed in detail and designs and
cost estimates produced. Sponish was recommended for prototype tidal development of the STO.

PROTOTYPE DEVELOPMENT

Prototype tidal development of the STO is planned at Sponish, North Uist,
Outer Hebrides (National Grid Reference NF 922694) in 1985 with commissioning in mid-1986. A single unit, three cell, STO barrage, with two independent vertical lift sluice gates each 2.5 x 2.5m square, is to be built
across the narrow entrance separating Loch Houram from Loch Maddy.

The STO operating head will vary between 0.5 and 3.0m and the flow passed
from 20-40m^3/s. It will generate electricity for supply to Lochmaddy Hospital in parallel with the North of Scotland Hydro-Electric Board system,
producing an annual output of 525MWh from an installed capacity of 270kW.
The energy will be used for heating, lighting and laundry facilities.

The barrage will be formed from reinforced concrete units laid on a prepared
foundation. The paddles, carriage and gates will be constructed from rolled
steel sections and plate which will be suitably treated to prevent corrosion.
The carriage will be articulated to allow for any slight barrage misalign-
ment and the paddles will have side and bottom edge rubbing strips to reduce
leakage. The pressure in the hydraulic circuit will be limited such that
standard fittings can be used throughout. The power take-off ram will be
compounded so that it provides equal resistance to the paddles in each direction, and will pump oil, via a bank of accumulators, to two variable displacement swash-plate hydraulic motors driving a single electric generator.
In the event of a failure or to allow maintenance, the STO can be dewatered
by stop-logging across the gates, so that the gates, paddles etc may be
readily removed.

The STO will be automatically operated by use of a micro-processor based
control unit, which will perform the necessary operational and safety
functions and optimize tidal cycle energy output by reference to the operating characteristics of Fig 3.

CONCLUSIONS

The STO offers a means of producing virtually inflation-proof energy at
ultra-low-head hydropwer sites throughout the world which were previously
considered to be uneconomic. Its simple design with low technology components make it particularly suitable for use in remote areas and developing
countries where often, the only alternative for energy production is high

TABLE 1 - SITES WITH POTENTIAL FOR DEVELOPMENT IN SCOTLAND USING THE STO

Site No.	Site	O.S. Map No.	National Grid Reference	Basin Area (km²)	Mean Tidal Range (m)	No. STO Units	Installed Capacity (MW)	Annual Energy Output (MWh)
12	Loch Gair	55	NR 931901	0.6	3.0	2	0.5	990
25	Loch Feochan	49	NM 822226	2.7	2.3	12	2.8	4,083
28	Upper Loch Creran	49	NM 980443	2.0	2.5	10	2.5	3,487
31	Upper Loch Leven	41	NN 136613	1.5	2.5	4	0.9	1,484
33	Loch Aline	49	NM 681445	2.0	2.9	10	2.6	4,011
34	Loch Don	49	NM 746315	1.3	2.3	6	1.5	1,940
40	Loch Teacuis	49	NM 623552	2.8	3.1	14	3.7	5,790
41	Upper Loch Sunart	40	NM 745610	6.5	3.1	24	6.4	10,650
43	Loch Ailort	40	NM 725791	5.5	3.2	24	6.5	10,150
44	Upper Loch Nevis	40	NM 801934	4.8	3.2	24	6.6	11,490
47	Loch Long	33	NS 881266	2.2	3.3	6	1.7	2,750
48	Loch Eishort	32	NG 635154	3.0	2.9	18	5.2	6,020
49	Loch Slapin	32	NG 570200	1.4	2.9	8	2.2	2,800
53	Acairseid Mhor (Rona)	24	NG 610564	0.15	3.6	2	0.5	953
54	Loch Sligachan	32	NG 530325	2.1	3.6	12	3.7	5,500
56	Loch Na Cairidh	32	NG 583290	1.8	3.6	10	3.0	4,610
59	Poll Creadha	24	NG 707413	0.5	3.6	2	0.5	1,000
62	Ob Mhellaidh	24	NG 829548	0.4	3.6	2	0.5	992
63	Badachro	19	NG 781746	0.3	3.3	1	0.27	420
67	Loch Roe	15	NC 059241	0.3	3.1	2	0.7	710
68	Loch Nedd	15	NC 130330	0.3	3.2	2	0.7	720
69	Loch Ardbhair	15	NC 163344	0.7	3.2	2	0.5	877
73	Loch a Chad-Fi	9	NC 210507	0.5	3.2	2	0.5	858
79	Loch Fleet	21	NH 805955	4.0	2.6	22	5.5	7,490
85	North Bay (Hoy)	7	HD 305912	3.9	2.3	16	3.8	5,220
87	Bagh Beag	31	NL 656978	0.15	2.5	1	0.3	280
91	Acairseid Mhor	31	NF 798096	0.15	2.5	1	0.3	281
93	Loch Eynort	22	NF 801273	2.5	2.8	9	2.3	3,240
94	Loch Skiport	22	NF 828387	0.4	3.1	2	0.6	780
95	Loch Sheilavaig	22	NF 849406	1.0	2.7	4	1.0	1,315
96	Loch Carnan	22	NF 832444	0.9	2.7	4	1.0	1,390
98	Loch Uiskevagh	22	NF 865510	3.1	2.7	12	3.0	4,352
101	Upper Loch Maddy	18	NF 914679	0.5	3.1	2	0.5	868
102	Loch Houram	18	NF 922694	0.3	3.1	1	0.27	525
103	Loch Portain	18	NF 939715	0.4	3.1	2	0.5	836
106	The Obba	18	NG 013866	0.2	2.9	1	0.27	380
109	Loch Beacravik	14	NG 114900	0.1	3.0	1	0.4	300
110	Loch Stockinish	14	NG 130908	1.0	3.0	4	1.0	1,587
113	Loch Mharabhig	14	NB 418199	0.5	3.0	4	1.4	1,257
114	Loch Erisort	14	NB 377215	7.6	3.0	28	7.4	11,430
115	Loch Leurbost	14	NB 383244	1.6	3.0	6	1.6	2,230
116	Loch Grimshader	14	NB 420258	0.8	3.0	4	1.4	1,238
117	Loch Meavaig	13	NB 092052	0.3	2.5	1	0.25	329
118	Loch Leosavay	13	NB 047076	0.2	2.8	1	0.25	383
122	Miavaig	13	NB 095340	0.14	2.9	1	0.27	354
123	Little Loch Roag	13	NB 130322	2.4	2.9	16	4.2	5,973
124	Loch Barraglom	13	NB 185335	1.6	2.9	6	1.6	2,296
125	Dubh Thob	13	NB 186364	0.2	2.9	1	0.26	362
126	Loch Ceann Hulavig	13	NB 211325	2.7	2.9	12	3.1	4,519

158   C. G. Carnie *et al.*

cost small-scale diesel generation.

The prototype development at Sponish will not only provide a valuable research tool, with data collected being used for the design of future developments, but its output will have a significant effect in reducing energy
costs at Lochmaddy Hospital.

REFERENCES

1. Report to the Scottish Development Agency and the Highlands and
    Islands Development Board, entitled "A study of potential tidal
    energy round the coast of Scotland", May 1984, by Wilson Energy
    Associates Ltd., Manchester and Crouch and Hogg, Consulting
    Engineers, Glasgow.

2. Two New Machines for hydraulic power from low heads - Wilson,
    Bassett and Jones, Symposium on Hydraulic Machinery in the
    energy-related industries, IAHR, Stirling 1984.

Tidal Streams and Energy Supply in the Channel Islands

P. R. Cave and E. M. Evans

Department of Mechanical Engineering, Plymouth
Polytechnic, Plymouth, UK

ABSTRACT

The Channel Islands illustrate a range of different economies all essentially
dependent on imported energy supplies. The current electricity supply systems
of the three major islands are described, and the economic case for investment in renewables, particularly tidal streams, is developed. The use of
cable links to Electricite de France is examined as these are in place or
under consideration in the islands. Current thinking on tidal stream energy
is examined, with a brief exploration of economics, technology, costs and
prospects. The islands are considered separately in detail, concluding that
Alderney has the best case for investment in tidal stream energy overall,
with interesting out of phase sites, Guernsey and Jersey could both make use
of the resource given a suitable investment climate, and that Jersey has a
very good site for a first prototype of economic size.

KEYWORDS

Tidal stream energy, economics, Alderney, Guernsey, Jersey.

INTRODUCTION

The work described in this paper forms part of a larger study into the
economics, technology and costs of tidal stream energy. This work has indicated a number of good sites for such generators in the Channel Islands.
These are essentially small independent states, dependent on imported energy
supplies, whose governments are able to take advantage of any energy supply
opportunity that is economically advantageous. The technical problems connected with the construction and operation of a tidal stream generator, or
tide mill, seem largely to have been overcome in recent years by work directed towards the offshore oil industry. There seems little doubt that a
tide mill could be built, installed and operated. The problem lies in the
economics: would such a mill be worth having? An attempt to indicate
answers to that question forms the body of the paper.

TIDAL STREAM ENERGY

Output and size

The output of a tide mill depends on the area of intercepted flow and the cube of stream velocity. Using values discussed more fully elsewhere by Cave and Evans (1984), and assuming uniform velocity distribution over a horizontal axis rotor, Fig.1 . is produced. For a typical site, mean output over a year is about a quarter of rated output.

Economics

The approach used is based on that developed by ETSU (1982), calculating permissible investment in a renewable replacing fuel. All cash values are corrected to mid-1984 using published indices (Monthly Bulletin of Statistics 1985). Assumptions made are a test discount rate of 5% and a plant factor of 0.3, which includes a load factor and availability. Plant lives beteween 10 and 25 years are considered, while three cases of fuel price rise are taken, with rises over a 25 year period of I : 200%, 2 : 100%, III : 50%. Costs for maintenance and transmission have been taken from ETSU (1982) by analogy with offshore wind generation.

Technology and installation

Each tide mill is envisaged as being of modest size (20 kW-1MW) and bottom mounted on a gravity structure. The general configuration proposed is shown in Fig.2.   The submerged rotor avoids the worst effects of waves. The electrical equipment would be mounted near the rotor with power transmission ashore by submarine cable. All the sites considered are near land and the cost of such a cable would not be excessive.

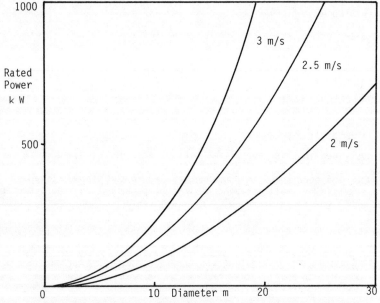

Fig. 1.   Size and rated output of a horizontal axis mill

Costs

A costing study using generic techniques is included in the programme at Plymouth, but results are not yet available. Miller (1983) gives the cost of a 4.3 m diameter horizontal axis machine designed to operate in 2 m/s flow as $1630/kW. This can be indexed to 1984 values to give £1245/kW. The assumptions made in the costing make it equivalent to "tide mill" values below.

Market

A number of sites have been identified in addition to the Channel Islands. They include the North Devon coast and locations in Scotland. There is a very promising site at Chindosudo in South Korea where 7 m/s streams occur.

CABLE LINKS TO EDF

A link to EDF is a current supply option in the Channel Islands. Jersey has just commissioned a link, while Alderney and Guernsey have them under consideration. The tariff is very complex (Pinta, 1984) with a fixed charge and metered structure that forces users to take electricity at off peak and peak times. Actual rates vary from 9.76 to 0.59 p/kWh, but neither of these is accessible in isolation. Rates averaging about 3 p/kWh seem to be achieved in practice. In addition to electricity costs, there will be cabling and connection to be borne by the user, and maintenance on the cable and standby system. This is needed as repair times on cable breakage could be 2-3 months. Jackson (1984) gave costs for a proposed connection from Alderney, with an energy demand in 1982 of 3.95 GWh including heating oil.

Average price of electricity	3.0 p/kWh
Capital charge for cable, 15 year life	3.5 p/kWh
Maintenance and standby system	5.2 p/kWh
Total cost at 3.95 GWh	11.7 p/kWh

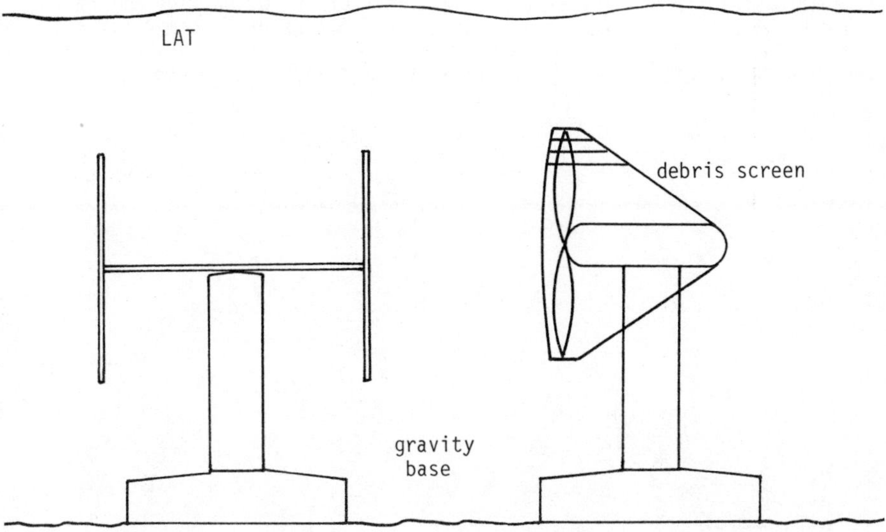

Fig. 2.   Possible configuration of tide mill

ALDERNEY

Current supply and economics

The current electricity supply on Alderney comprises five diesel engines all using distillate fuel.Installed capacity is 1.52 MW, maximum demand to 1984 was 980 kW, the minimum was 220 kW.The cost of production in 1982 (Jackson, 1984) was 10 p/kWh, and total production was 3.57 GWh. Using the economic arguments outlined above, and with a 1984 fuel value of 4.79 p/kWh the case for investment in tidal stream generators is given in Table 1.

Tidal Flows

A good description of the tidal flow round Alderney is given by Warburg (1945). This description confirmed by limited survey and local experience, is summarised in Fig.3.   The main flows, exceeding 3.5 m/s in the Swinge and Race, change direction together on a normal tidal cycle. The two eddy flows are asymmetrical in time and almost exactly out of phase so that slack water to the south coincides with maximum flow to the north. A pair of sites could thus be found, physically close to each other, out of phase, and with stream velocities up to 2 m/s. These would make possible a contribution to firm power, favourably altering the economic case.

Discussion

Alderney currently has a reliable electricity supply which meets the needs of its consumers. It is, however, expensive. Energy supply on Alderney is currently under review (Jackson,1984) and a case for a link to EDF has been made. This proves to be uneconomic unless demand rises to 10 GWh, and this depends on economic growth which may not be possible. The case for tidal energy is a good one. On the smallest projected rise in oil prices (III) investment is worthwhile over 20 years. With the largest rise (I), payback should be possible over about 12 years. The tidal regime makes possible a contribution to firm power, and an installation on this basis could be sized always to meet the minimum demand of 220 kW. Surplus production at night could be absorbed by water heating using load management techniques, and used usefully during the day. Overall, Alderney presents a very good case for more detailed study into the utilisation of tidal stream energy.

Fig. 3.   Tidal streams around Alderney

GUERNSEY

Current supply and economics

The current electricity supply on Guernsey consists of 3 Sulzer diesels
rated at 11 MW each and 6 Mirrlees diesels rated at 3.5 to 3.8 MW each. All
these burn heavy fuel oil. There is also a gas turbine rated at 11 MW. The
maximum demand to 1983 was 43.5 MW, the minimum 8 MW. Total consumption in
1982/3 was 167 GWh. The cost of fuel in 1984 is taken as 3.0 p/kWh, giving
the results in Table 1.

Tidal flows

Guernsey is in the main stream flowing along the French coast. Sites can be
found all round the island where 2.5 m/s is reached, but no detailed survey
has been made. Warburg (1945) reports an out of phase effect where the
stream along the north coast is turning while the main stream is flowing south
westward at full strength.

Discussion

Energy supply has recently come under review, and a link to EDF is being
investigated. Stability problems caused by its length may necessitate a DC
link, increasing costs, however. The case for tidal stream energy based on
fuel replacement can only be made if the worst case fuel price rise (I) is
combined with a 25 year view. If a pair of out of phase sites could be located, the economic argument would alter favourably.

JERSEY

Current supply and economics

Jersey has a complex energy supply system based mainly on a 90 MW heavy oil
burning steam turbine plant. There are also 4 distillate fuel diesels rated
at 5 MW each and a 17.5 MW gas turbine. The Resources Recovery Board (RRB)
operates a refuse burning plant producing 3 MW of steam and a 1.7 MW diesel
plant burning refuse derived methane. About half the RRB output is sold to
Jersey Electricity. A 40 MW link to EDF has just been commissioned. The domestic tariff in 1984 was 6.7 p/kWh and the purchase rate from RRB was

Payback Years		Case I £/kW	p/kWh	Case II £/kW	p/kWh	Case III £/kW	p/kWh	
10	total	1369	6.7	1170	5.77	1071	5.28	Alderney
	tide mill	896		699		600		
15	total	2049	7.5	1678	6.15	1492	5.47	Alderney
	tide mill	1526		1155		969		
20	total	2686	8.2	2128	6.5	1848	5.6	Alderney
	tide mill	2092		1534		1254		
25	total	3269	8.8	2521	6.8	2148	5.8	Alderney
	tide mill	2614		1853		1493		
20	total	1693	5.12	1342	4.1	1165	3.53	Guernsey,Jersey
	tide mill	1099		748		571		
25	total	2060	5.56	1589	4.29	1354	3.66	Guernsey,Jersey
	tide mill	1405		934		699		

Table 1. Fuel replacement economics.

E3.5

3.34 p/kW, geared to oil prices. Maximum demand to 1985 was 97 MW and the
minimum 14.6 MW. Total production in 1982/3 was 327.5 GWh. The cost of
fuel in 1984 was taken 3.0 p/kWh giving the results in Table 1.

Tidal Flows

Jersey is out of the main flow along the French coast, but does have a
larger tidal range than the other islands. Warburg (1945) reports that slack
water occurs on north and south coasts with full flow along east and west
coasts, and vice versa. A number of sites exist off all coasts where 2.5 m/s
occurs and one of these has been surveyed in detail.

Discussion

Jersey has a robust power supply system which includes a link to EDF. This
link will never meet full consumption and a standby will have to be kept
anyway. Jersey Electricity is used to obtaining power from unconventional
sources and means to reduce import costs are always of interest. The case
for tidal stream energy requires a long view, but in the shorter term there
is a very good site for a prototype tide mill cantilevered out from St.
Catherine's breakwater into the tide race. This could be a small (4m) but
useful (10 kW) device and would give valuable operating experience.

ACKNOWLEDGEMENTS

The authors acknowledge support from SERC for the work described here, from
Alderney Electricity Ltd, the States of Guernsey Electricity Board and the
Jersey Electricity Company in the provision of information, and Mr. A.W.
Carter of RRB, Mr. Ron Smart of Alderney and Mr. Trevor Bull of Jersey for
much useful information.

REFERENCES

Cave, P.R. and Evans, E.M. (1984). Tidal stream energy systems for isolated
	communities. Alternative Energy Systems, Pergamon, 9-14
E.T.S.U. (1982) A strategic review of the renewable energy technologies.
	E.T.S.U. R13, H.M.S.O.
Jackson, R.J. (1984). A consultancy study. Alderney Electricity Ltd.
Miller, G., Corren, D., Franchesci, J. and Armstrong, P. (1983). Kinetic
	hydro energy conversion schemes and the New York State resource.
	NYU/DAS 83-108, New York University.
Monthly Bulletin of Statistics (1985). Vol XXXIX, no. 4, April 1985,
	Statistical Office, U.N., New York, 198-199.
Pinta, J.C. (1984). Electricity tariff structures for supplies above 36 kW,
	period 1984-1990. OA Trans 2684, Electricity Council, 1985.
Warburg, H.D. (1945). Tidal streams of the waters surrounding the British
	Islands and off the West and North Coasts of Europe. Hydrographic Dept.
	of the Admiralty, London, 68-75.

Microprocessor Control of Village Hydroelectric Systems

P. Robinson

Department of Electrical and Electronic Engineering,
Plymouth Polytechnic, Plymouth, UK

ABSTRACT

An inexpensive, on-site programmable, single chip microprocessor controller is described including details of interface circuits. Experience gained between 1980-1984 running an electronically controlled 7.5 kW hydroelectric system in a remote Papua New Guinea village is discussed. Particular emphasis is given to installation, control and maintenance problems.

KEYWORDS

Microhydro control; microprocessor control; hydroelectric systems; Papua New Guinea.

INTRODUCTION

An electronic load controller or governor is a device which is designed to ensure that the electrical load on a generator remains constant. If this condition, i.e. constant load, can be achieved then the cost of a microhydro system may be substantially reduced. Savings of up to about 30% of the total capital cost are possible. This is because there is now no need for the expensive hydraulic system which controls the flow of water into the turbine. The hydraulic system consists of a controller which responds to changes in the electrical load on the generator by adjusting a high pressure water valve. This valve ensures that the water flow into the turbine is just sufficient to maintain the required output power from the generator. Unfortunately this valve and the associated control gear are usually the most expensive pieces of capital equipment in a microhydro system.

In remote areas of Papua New Guinea the typical microhydro system is a simple run-of-the-river installation. A small weir is built across a stream to provide some settling of silt, leaves, etc. A pipe is then run from this settling area down the mountainside to the generator. In many instances there are no valves on the pipe and the generator is run at full power continuously. A large fixed load such as a water heater is permanently wired to the generator and the variations in consumer load cause the system voltage and frequency to vary. At night when lights are switched on both the frequency and voltage fall. These simple systems will only operate in a satisfactory manner if the variable load is small compared to the fixed load. In practice this usually means that only a few consumers may be connected to the system, although the large fixed load could be for communal use, e.g. water heating. A more sophisticated arrangement is necessary for a village system.

If the load on the generator is to remain constant then there must be some means of measuring the consumer load and then adding it to a secondary load so that the total remains constant. The differences between various electronic load controllers lie in the methods used to achieve these requirements; i.e. measurement and load adjustment. On the assumption that the water flow into the turbine remains constant then an increase in electrical load on the generator causes two easily observable effects. These are a decrease in both terminal voltage and drive shaft speed. Either may be used as an indication of the system load. Over recent years the drive shaft speed has been the preferred method of indicating the system load. This is because the frequency of the generator output voltage, which is directly proportional to the shaft speed, may be easily measured using digital techniques. In addition the frequency is constant throughout the system thus enabling the measurement to be taken at the most convenient point. Reliable voltage measurements are usually restricted to the generator output terminals.

THE CONTROLLER

One of the main design criteria for the microprocessor controller was that it should have a minimum number of components. This led to the choice of a R6511AQ chip at the heart of the system. This chip includes a 6502 C.P.U., 192 bytes of RAM, 32 bidirectional I/O lines, two 16-bit programmable counter timers plus many other features. The need for interface and RAM chips was thus eliminated. Figure 1 gives an overview of the complete system.

The input and output circuits interface the controller directly to the 240 volt mains distribution system. The main program resides in a single EPROM chip whereas on-site adjustment is implemented by switches which use the RAM, available on-board the R6511AQ for storing these variables. Power may be supplied from battery backup but in this case the controller derived its power from the generator.

Controller operation may be best described by considering a change in the electrical load on the generator. If the load is increased then the shaft speed decreases and the system frequency falls from say 50Hz to 49Hz. By means of the input circuit the controller detects this reduction in frequency and reduces a variable, controlled load such that the frequency returns to 50Hz. This controlled load, often known as the dump load, is usually a communal water heating system.

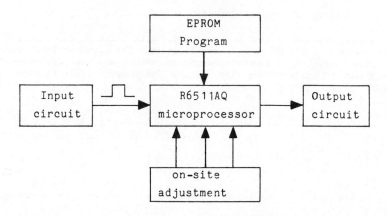

Fig. 1 System block diagram

There are many problems to be overcome to make such a system a
practical proposition.   Firstly it is impossible to maintain the
frequency at exactly 50Hz.   Therefore it is necessary to establish
some frequency band, centred on 50Hz, within which the frequency will
be allowed to vary without precipitating load changes. This is called
the dead-band.   Typical dead-bands are from +/- 0.25Hz to about +/-
1.5Hz.   Only when the frequency has moved outside these pre-determined
limits will the controller implement a load change to correct the
deviation.   System time constants present further problems. At 50Hz
the periodic time, i.e. the time for one revolution of the generator
shaft, is 20ms. The microprocessor completes its calculations in less
than 1ms whereas transients due to the mechanical inertia of the
turbine/generator combination may last for many hundreds of
milliseconds.   It is clear that the frequency measurements made
directly after a load change may be invalid due to the presence of
system transients.   These problems are different in magnitude for
nearly all hydro systems.

A major advantage of the microprocessor controller over its
predecessors is its ability to cope with the variations of system
parameters from one scheme to another. On-site adjustments enable the
installer to optimise the response of the controller to ideally match
that of the particular microhydro system. These adjustments are wait
delay, width of dead-band and integration period. The wait delay is
designed to prevent load switching until all transients from the
previous load change have died away.   The width of dead-band is
specific to the particular system and depends upon the droop
characteristic of the generator and the minimum dump load available.
For example, if the frequency of the generator falls by 1.0Hz for each
ampere increase in load current and the minimum dump load available is
1.0A then the minimum dead-band must be 1.0Hz, i.e. 50 +/- 0.5Hz. If
a narrower dead-band is used then the system is likely to oscillate.
This is because a minimum dump load change is sufficient to cause the
output frequency to move from below the minimum dead-band limit to
above the maximum dead-band limit.   Consequently this minimum load
will be switched in and out of the system continuously. The final
adjustment allows the periodic time to be averaged over several cycles
in order to increase accuracy.

INPUT AND OUTPUT CIRCUITS

The input circuit converts the 240 volt a.c. mains signal into a TTL
level square wave of the same frequency. The edges of this square
wave are used to interrupt the microprocessor and therefore provide
the means of determining the system frequency. This is shown in
Fig.2.

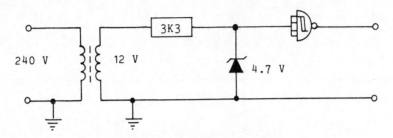

Fig.2. The input circuit

Four independent output circuits were used at Baindoang village. Each
output circuit operated a different value of dump load. These loads
were 0.5, 1.0, 2.0 and 4.0 kW giving sixteen possible load
combinations to a maximum of 7.5kW. The binary weighting of the loads
was useful but not necessary. In this case the slope of the generator
droop characteristic over the critical region of between 49 and 51Hz
was found to be - 0.8 Hz/A. If the mains voltage is assumed to be 250
volts then the current through a 0.5 kW load is 2A giving the minimum
allowable dead-band of 1.6 Hz, i.e. 50 +/- 0.8Hz. In practice the
dead-band would be set a little wider in order to provide a margin of
safety.

Details of a single output circuit are given in Fig.3. An output pin
of the microprocessor is connected directly to the input of this
circuit. To switch this load on the microprocessor output must go
high, i.e. + 5 volts dc. This is inverted by the 7416 open collector
inverter causing a current to flow from the dc supply through the
light-emitting diodes to earth. Two of these diodes activate triacs
in the gate circuits of the power thyristors. The back to back
thyristors are then switched on by natural commutation. The third
light emitting diode is merely an indicator showing which loads are on
at any particular time. Natural commutation ensures that all load
switching is done near the zero crossing of the voltage waveform. The
spikes normally associated with thyristor switching are therefore
nearly eliminated. An interesting feature of this system is the
absence of A/D and D/A convertors. A further advantage of this
control system is the absence of any synchronisation between the input
and output signals. Complicated, and therefore potentially
troublesome, timing circuits are no longer necessary. At Baindoang
village these four loads were located within a large steel drum and
provided a constant supply of hot water to two communal showers.
Robinson and Kormilo (1984) include a more comprehensive description of this
system.

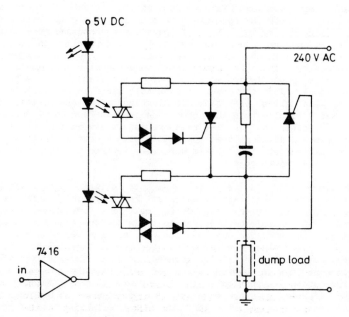

Fig.3. A single output circuit.

INSTALLATION AND MAINTENANCE

The Baindoang hydroelectric system was initiated by Mr. J. Yang, a teacher at the village community school. In September 1975 he heard a local radio programme about the potential for village hydroelectric systems and promptly wrote to the Papua New Guinea University of Technology requesting assistance in developing such a system for the Baindoang area. Most of the civil and building work was completed between November 1976 and June 1979. This was a self help scheme with the villagers providing labour, the University the technical expertise and funding from a variety of sources. A thorough and fascinating account of this work is given by Inversin (1981).

The author first visited the scheme in 1980. At that time there was no effective system control. A large heating load was permanently wired to the generator and a few small fluorescent lights wired in some of the teachers' houses. It was apparent that as the house wiring was extended there would have to be a suitable load controller. This need led to the development of several controllers culminating in the successful programmable microprocessor based system. The final version of this controller was installed on site in mid 1984 and, as far as is known, is operating in a satisfactory manner.

By the end of 1980 all the 'agents of change' who had been active in the design and implementation of the Baindoang system had left. This included the dynamic Mr. Yang. Community school teachers in Papua New Guinea, as a matter of national policy, rarely teach in their own home

areas and tend to move fairly frequently. The villagers were
therefore left with the formidable problem of running and maintaining
a fairly complex system. None of the villagers had had more than an
extremely elementary education and technical expertise was virtually
non-existent. A further problem was that the village elders, who had
some authority in the community, were totally unsuitable for
organising or supervising maintenance. In the end the only teacher
who was local, and therefore permanently based in the village, Mr.
Yunangmor, was given the task of looking after the system. This
teacher was the oldest in the school and looked after the first grade
class. The other four or five teachers rarely stayed for more than a
year or two and although they appreciated the electric lights and hot
showers provided by the system were not prepared to become actively
involved in its maintenance.

Mr. Yunangmor did his very best as the official system manager.
Unfortunately both his lack of technical knowledge and his fairly low
standing among his peers tended to make his position extremely
difficult. An example of his technical limitations, and the author's
presumption, is provided by an account of the worn out turbine
bearings. Each time the author visited the village the turbine and
generator were examined. The grease nipples on the bearings of the
Gilks turbine always showed signs of being freshly greased and the
grease-gun stood reassuringly in the corner. However, bearing noise
increased alarmingly with each visit. Not until the bearings were
close to destruction was it discovered that a small quantity of grease
was regularly pumped from the gun on to a finger and then carefully
placed on top of the grease nipple!

Further problems arose over the collection of rents. Initially the
teachers were the only ones with lights inside their houses. They
were paid a regular salary from the government and were expected to
make monthly payments which would help to pay the running costs of the
system. Unfortunately many refused to do so. The ultimate penalty was
that the offender be disconnected. Not unnaturally Mr. Yunangmor, the
most junior of the teaching staff, was rather reluctant to face the
wrath of his head master which would result from a disconnection.
Because of this, and other factors, the income from the system was
irregular and did not meet the true running costs.

Between 1980-1984 the Papua New Guinea University of Technology
subsidised the maintenance and further development of the system.
Without this contribution there is no doubt that the system would have
ceased working. A full account of this work can be found in Robinson
(1984). From the University point of view this involvement enabled
many students to gain valuable practical field experience. It also
provided an excellent testing facility for various areas of research.

REFERENCES

Robinson, P., and S. Kormilo (1984). Proc. First International
    Conference on Small Hydro, Singapore.
Inversin, A.R., (1981). Technical Notes on the Baindoang Micro-Hydro
    and Water Supply Scheme, ATDI, Private Mail Bag, Lae, P.N.G.
Robinson, P., (1984). Baindoang Microhydro Project 1980-84, Dept. of
    Elec. & Comm. Eng., Private Mail Bag, Lae, P.N.G.

A Low Head Hydro Scheme Suitable for Small Tidal and River Applications

P. R. S. White, L. J. Duckers,
F. P. Lockett, B. W. Loughridge,
A. M. Peatfield and M. J. West

Energy Systems Group, Coventry (Lanchester) Polytechnic,
Coventry, UK

ABSTRACT

The need to provide small communities with electrical power is well
recognised. In many cases these communities are close to rivers or small
estuaries which have energy capacities in the 10 kW to 100 kW range, and
with an available head of 1 m to 2 m which is considered low in
conventional hydro terms. The paper describes the operation of a modular
device consisting of a chamber which acts as a water to air gearbox as
water enters and leaves so expelling and inhaling air into and out of the
chamber via a Wells turbine. Model tests of device performance are
reported, and are used as input to the simulation of a full scale
turbogenerator. Outline full scale design and costings are also discussed.

KEYWORDS

Low head; micro hydro; Wells turbine; hydro electric; river application;
tidal application.

INTRODUCTION

The generation of small amounts of electrical power is of interest from two
main points of view. In the U.K. since the 1983 Energy Act which obliges
the C.E.G.B. to buy privately generated electricity of a technically
acceptable quality, a small scheme can generate income. For example a
scheme which produces an average of 100 kWe over the year would yield about
£20 000 per annum when the variable buy in tarriffs are taken into account.
Secondly where small communities have been relying on diesel driven
generators or have not previously been connected to a supply there is a
great need to provide electrical power at an economic rate, which when
social factors have been taken into account may be different to that
required when selling power to the C.E.G.B. or other similar utility. In
many cases these communities are close to rivers or small estuaries which
have energy capacities in the 10 kW to 100 kW range, but with an available
head which is in the range of 1 m to 2 m. When conventional hydro systems
are considered these heads are too low for the economic utilisation of
equipment.

The paper is concerned with the principle of operation, model testing, adaptation to full scale with consideration of costs and quality of electrical output of a modular device which can operate successfully with heads as low as 1 m, and which is applicable to both river and tidal applications.

PRINCIPLE OF OPERATION

In its simplest form the system consists of an enclosed chamber into which the available water flow can be controlled by the operation of inlet and outlet valves in such a way that the effective driving pressures for both the filling and the emptying cycles can be a large proportion of the head available at the site chosen.

At the top of the enclosed chamber, above the maximum height of the upstream water level, is an inlet-outlet air duct leading through an air turbine to atmosphere, see Fig. 1. In low head hydro applications the Wells turbine (White, 1981) would be well suited to provide a power take-off unit with a high rotational speed suitable for electrical generation, with its ability to operate in reversing flow without the use of rectifying valves being a major advantage.

The operational cycle commences with the opening of the water inlet valve, allowing water into the empty chamber with the outlet valve closed, thus filling the chamber with water and driving the air out under pressure through the rotating Wells turbine. At a suitable point near the end of the filling cycle the inlet valve is closed and the outlet valve is then opened allowing exit of the water to the downstream side of the water retaining structure. This emptying process causes air to be sucked back into the chamber through the still rotating Wells turbine which is thus used to extract energy at the optimum rate during both parts of the cycle. Then at a suitable point near the bottom of the emptying cycle the outlet valve is closed and the whole cycle recommences with the re-opening of the inlet valve. Full details of the operation and control of the device are given in Peatfield et al (1984).

MODEL TESTS

A 1/5th linear scale model of a device to operate on a 20 kW resource is being constructed in the laboratory (Fig.2), and consists of a wooden box with a cross sectional area A (Fig.1) of 0.67 m^2 . The valves are pneumatically operated, and are of the vane damper type operating in a rectangular orifice with an area chosen to minimise hydraulic losses. The performance of the turbine is simulated using a rotating damper (White et al, 1985) which reproduces the linear relationship between flow rate and induced pressure drop characteristic of the Wells turbine, and has the added advantage that a change in the operating speed of the damper will change the damping presented to the air, and so permit rapid optimisation of the model.

An initial model to verify the concept was constructed for inclusion in a narrow (0.3 m) water channel. Figure 3 shows a trace of the pressure drop across the damper which was obtained using manually operated vertically sliding gate valves to control the flow. The peaks on the trace are associated with the rapid movement of water in the box at the ends of the operating cycle. It is thought that this results from the large length to width ratio (5) of the box which is a direct result of using the narrow water channel. It is not anticipated that this will occur in the current model.

FULL SCALE ADAPTATION

The box for a unit operating on a 20 kW resource with an available head of 2 m would have a cross sectional area of approximately 17 m^2 and a valve area of 1.9 m^2. The corresponding cycle time and water velocity through the valves would be 20 s and 0.5 m/s respectively. Full details of the cycle analysis are given in Peatfield et al (1984).

The corresponding turbine is a 0.5 m diameter two plane rotor Wells turbine rotating at 6 000 rev/min and driving a 20 kW induction generator at 3 000 rev/min. Based on costing experience gained in the wave energy programme the estimated costs are :-

> concrete box   £ 5 000
> turbogenerator and valves   £ 5 500

which corresponds to £ 525 per installed kW.

The expected electrical output was derived from a simulation of the turbine and generator performance with a sinusoidal pressure variation across the turbine, which has a similar effect on the turbine to that obtained from a model test and shown in Fig.3. Figures 4a and 4b show the electrical output, and the effect on quality achieved by increasing the effective inertia of the turbogenerator. The degree of smoothness, and therefore the added inertia required, depends on the use to which the supply is to be put.

The system described requires the construction of a box which can be located in a river, leat, or small estuary so providing power from river and tidal situations. With tidal uses the box can be used for two way generation by means of valve sequencing.

DISCUSSION

The modular system described has a low cost per installed kW and can be tailored to suit many tidal and river locations. The main feature is in the use of the box and valves as a water to air gearbox which permits the use of a low cost air turbine which can operate at high efficiency with heads in the 1 m to 2 m range, in place of a water turbine, which is typically upto six times more expensive per kW than the air turbine proposed. A further advantage in using an air turbine as opposed to a water turbine is that the turbine house is above water level. This reduces construction costs as the turbine does not have to be located well below water level to reduce cavitation effects, and also makes maintenance access easier.

Already through the British Technology Group commercial interest has been shown, and the next stage in the project is to determine the operating efficiency of the box under different operating conditions of flow, head, stroke and damping rate.

174   P. R. S. White *et al.*

Whilst it is appreciated that the major weakness of the system is the
mechanically operated valves the design of pneumatic self acting valves is
in hand, as is the development of other types of water to air gearbox for
providing the air flow to a turbine, again with an available head in the
region of 2 m.

With the work described above, and that currently in hand the future for
micro head hydro schemes for river or tidal applications looks very
promising for both U.K. and overseas requirements.

ACKNOWLEDGEMENTS

The authors would like to acknowledge the support of Ocean Energy Ltd., and
Coventry (Lanchester) Polytechnic for providing the necessary facilities.

REFERENCES

Peatfield, A.M., Duckers, L.J., Lockett, F.P., Loughridge, B.W.,
    White, P.R.S., West, M.J., (1984). Energy from low head water sources.
    Alternative energy systems. Electrical integration and utilisation.
    Pergamon Press pp 1-8.
White, P.R.S. (1981). The development and testing of a 1/10th scale
    self rectifying air turbine power conversion system. W.E.S.C.
White, P.R.S., West, M.J., Duckers, L.J., Lockett, F.P., Loughridge, B.W.,
    Peatfield, A.M. (1985). The design instrumentation and use of linear
    orifices to simulate the performance of Wells turbines. Fourth
    International Conference on Systems Engineering. Coventry (Lanchester)
    Polytechnic.

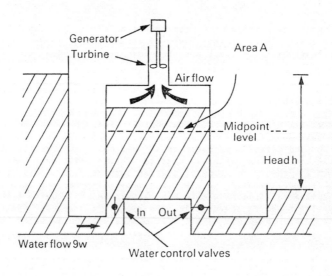

Fig. 1. Schematic diagram of chamber.

Fig. 2. 1/5th linear scale model.

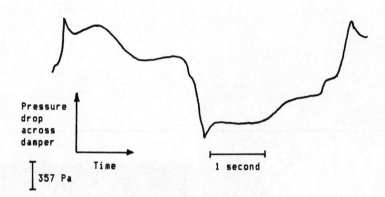

Fig. 3. Pressure trace across damper.

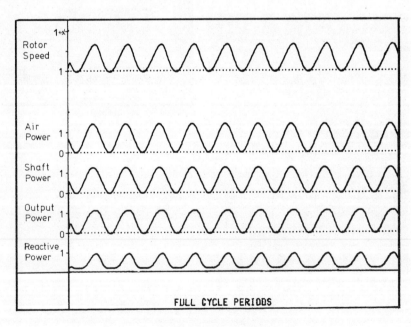

Fig. 4a. Low inertia performance.

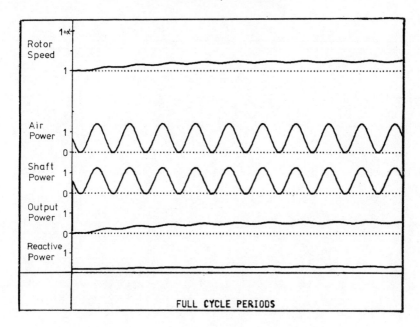

Fig. 4b. High inertia performance.

Tapered Channel Wave Power Plants

A. E. Fredriksen

Universitet 1 Trondheim, and Norwave A/S,
Forkningsveien 1, Blindera, Oslo 3, Norway

KEYWORDS   Wave power, tapered channel focusing breakwater

ABSTRACT

In many parts of the world the energy content of the ocean waves is considerable and the cost of traditional electricity is high.   This makes wave power
an interesting proposition.   The critical problem is the development of
efficient and reliable methods for converting wave energy into electrical
energy.   In general, such energy conversion devices must meet the following
requirements:

*   The conversion efficiency should be insensitive to variations in
     wave height, frequency and direction.

*   The construction itself should be able to cope with extreme weather
     conditions.

Most of the devices proposed for wave energy conversion involve the exposure
of moving parts and machinery to adverse off-shore conditions.   As a result,
production and maintenance costs for these devices are high.

To overcome these problems, Norwave A.S. and the Central Institute for
Industrial Research, Oslo, Norway, have developed a novel method for wave
energy conversion: the tapered channel wave power plant (Tapchan).   Tapchan
Power Plants are now offered by Norwave A.S. on a commercial basis.

1.   THE TAPCHAN POWER PLANT : INTRODUCTION

The Tapchan Power Plant combines the following features:

*   The wave energy is converted to potential energy in an on-shore water
     reservoir.

*   The generation of electricity is carried out by standard hydroelectric
     power plant technology.

*   The conversion device is entirely passive and has no moving parts.

*   High energy conversion efficiency is maintained over a broad range of
     wave heights, frequencies and directions.

E6.1

The principle of operation of the Norwave Tapchan can be explained by dividing the system into the following sub-systems: (see fig.1)

*   A collector designed to concentrate the wave energy and optimize collectio
     efficiency for a range of frequencies and directions.

*   An energy converter in which the energy of the collected waves is
     transformed into potential energy in a water reservoir.

*   A water reservoir for storage of the converted energy.   The reservoir
     also serves to average out short-term fluctutations in the wave energy
     level.

*   A conventional hydroelectric power plant for electricity generation.

(a)   The wave energy converter

This stage is the unique part of the Tapchan Power plant.   It consists of a
gradually narrowing channel with wall heights equal to the filling level of
the reservoir (typical heights 3-7m).   The waves enter the wide end of the
channel and, as they propagate down the narrowing channel, the wave height is
amplified until the wavecrests spill over the walls.   The gradual narrowing
of the channel causes a continuous sideways spill-off as the wavecrests move
along.   As a result, the wave energy is gradually transformed into potential
energy in the reservoir.   The advantage of this conversion method is that it
is entirely passive, and almost completely insensitive to variations in wave
height and frequency.

Various models of converters have been thoroughly tested in a wave flume as
well as in a wave tank and shown to behave according to specifications.
Typical figures for the conversion efficiency are in the range of 65-75%.

(b)   The collector

The collector serves the dual purpose of concentrating the incoming waves
before they enter the converter, and matching the converter to the local wave
climate in an optimum manner.   The horn shaped collector is designed so
as to optimally collect the wave energy over a range of incoming wave
frequencies and directions.

In some cases it is desirable to modify the local wave climate (i.e. increase
the energy density) by using offshore lens elements for wave focusing.   Such
wave lenses act as a pre-collector, and must be adapted to the collector
stage.

(c)   The reservoir

The main function of the reservoir is to provide a stable water supply for
the turbines.   The reservoir must be large enough to smooth out the fluctuatio
in water spill-off due to the individual variations in the waves propagating
down the tapered channel.   This is achieved with a small reservoir.
Storage capacity for less than one hour's operation of the turbines is ample.
The reservoir should be as small as possible, the lower limit being determined
by the technical requirements of the water turbines.

(d)   The hydroelectric power plant

Well established techniques are used for the generation of electric power.

                                                                    E6.2

The water turbine driving the electric generator is of a low head type such as a Kaplan, or a tubular turbine. It must be designed for salt water operation, and should have good regulation capabilities. Otherwise, there are no special requirements for the power station and its eqipment.

2.   DESIGN OF A TAPCHAN POWER PLANT

The Tapchan power plant is based on a concept which is adaptable to a large spectrum of external conditions. Plants of sizes between 0.5 MW and 300 MW can be constructed.

A Tapchan power plant is individually designed to ensure an optimal configuration. The capital costs of the plant and the unit cost of power, i.e. average production cost per kWh in a year, should be minimised.

Normally Norwave will start with a feasibility study in order to evaluate the possibilities for an economical utilization of the available wave power in a region. The economy of any wave power plant is highly dependent upon local variations in the wave energy. Key parameters, such as capital cost per installed kW and unit cost, vary with installed machine capacity, dimensions and shape of converter etc. Thus, optimization of plant parameters is a very important part of the design.

Norwave and partners have developed the theoretical basis and adequate design tools for optimization, design, and performance analysis of Tapchan power plants. With our Wavetrack computer program, the local wave climate is analysed. The Wavetrack results are used together with results from well established procedures for site selection to determine the most advantageous plant location.

Computer programs are important tools in the optimization and design of the power plant and its sub-systems. One example is a program for design of optimum size and shape of the collector stage of the system. With a given spectrum of sea waves as input, this program calculates the detailed wave patterns and the overall transmission efficiency of converters of various shapes. The output data from the program are important parameters used in a computer-aided optimization procedure in which plant parameters (dimensions of converter and reservoir, installed machine capacity, etc.) are determined to give optimum overall performance of the Tapchan power plant.

Norwave has found that great cost reductions can be achieved by a thorough and intelligent optimization and design. Optimization of potential Tapchan power plants has resulted in reductions in unit costs of power by factors of between 2 and 4. The unit cost of power from some actual Tapchan power plant propositions has been calculated to less than 2 new pence per kWh (about 3 U.S. cents).

This figure compares favourably with most other energy costs, and indicates that our Tapchan wave power plant in many cases might be the right device for the supply of electric power.

3.   PARTICIPATION BY NORWAVE A.S. IN WAVE POWER DEVELOPMENT

Norwave A.S. is a company formed by former staff members of the Central Institute for Industrial Research, (SI), to develop and market products based on wave phenomena. The Norwave employees and their SI colleagues have carried out a major research programme on ocean waves since early in the seventies. One notable result is the renowned focusing wave power plant, in

which specially designed water wave lenses are used to convey the wave energy
into a focal area, where the converter is placed. This concept is well
suited for large wave power plants, producing several hundred megawatts.

During the later years SI and Norwave A.S. have concentrated on the
development of smaller wave power plants. The result of this effort is the
Tapchan Wave Power Plant which now can be constructed on a commercial
basis.

Norwave A.S. and its partners in Norwegian industry combine some of the
world's most advanced scientific expertise on wave power, with extensive
international experience in the design and construction of hydroelectric
power plants, as well as large offshore structures.

Norwave A.S. can assist in connection with any activity related to the
development of wave power.   This assistance can range from cooperation with
local companies on special activities, to taking full responsibiliity
for any part of the design and construction of a Tapchan Power Plant. This
includes turnkey delivery of the complete power plant.

For further information and references, please contact.

Mr Thomas Hysing, General Manager
NORWAVE A.S.
Forskningsveien 1,
Oslo 3
Norway
Telephone: 45 20 10
Telex : 71536

Fig.1.   The tapered channel breakwater wave power system

TOPIC F

Wind Power

Wind Monitoring for Large Scale Power Generation on Shetland

J. A. Halliday[*,**] P. Gardner[**] and E. A. Bossanyi[*]

*Energy Research Group, Rutherford Appleton
Laboratory, Didcot OX11 0QX, UK
**Energy Studies Unit, University of Strathclyde,
Glasgow G1 1XQ, UK

ABSTRACT

The paper describes the establishment of 2 hill top monitoring stations in
Shetland and highlights the practical problems encountered and the
solutions found. The results obtained to date are briefly summarised.

KEYWORDS

Wind; Monitoring; Shetland; Diesel.

INTRODUCTION

In September 1984 a 3 year collaborative project commenced, involving the
Energy Research Group of the Rutherford Appleton Laboratory (RAL); the
Energy Studies Unit of the University of Strathclyde; and the North of
Scotland Hydro Electric Board. It has the overall aim of assessing the
potential for large scale power generation on Shetland using wind turbines.
This assessment has 3 parts - (1) To assess the long-term wind potential of
two hill tops as sites for large (rated power more than 200 kW) wind
turbines, by (i) monitoring the wind regime at the two hill tops, (ii)
comparing the measurements made with those made simultaneously at two
established Meteorological Office stations and (iii) using historic records
to infer the long-term wind regime of the hill top sites. (2) To assess
the savings attributable to wind energy and to identify the changes which
would be required in the operation of the existing generation plant in the
island's power station. This assessment will make use of a timestep
computer simulation model (which will use not only the wind data monitored
at the hill top sites but also load data and a detailed knowledge of the
operating characteristics of the existing generation plant). (3) To assess
whether the presence of wind turbines connected into the island electricity
grid will cause changes to the electrical quality and adversely affect the
operation of the existing generation plant.

This paper will only describe how the hill top monitoring experiment was
planned and executed - details of other parts of the project, such as the
time step simulation model, will be described in future papers and reports

SHETLAND

The Shetland group consists of over 100 islands (with a total area of 1,408 square kilometres), situated to the north of Scotland (See Fig. 1). Most of the population of 22,830 (estimated at 31 December 1983), (Shetland Island Council, 1983), live on the largest island, Mainland, which has an area of 899 square kilometres (347 square miles). The economy of Shetland is dominated by the North Sea oil industry, though the completion of the Sullom Voe oil terminal has led to an increase in unemployment, and a decrease in the resident population. The electricity consumption on the islands has undergone a remarkable change since 1971 (units generated = 32 TWh, maximum demand 11.69 MW), rising to a peak in 1982/83 (units generated = 146 TWh, maximum demand 32.30 MW) and more recently declining 1983/84 (units generated = 132 TWh, maximum demand = 28.90 MW). The total number of consumers connected to the island grid in 1983 was 10,453.

The two hill tops being monitored were chosen by the Hydro Board after consultation with the island authorities, the land users and owners and consideration of relevant factors such as the proximity of existing power lines, access, topography and the existing use of the hills. The locations of the 2 hills - Scroo Hill and the Hill of Susetter - are shown in Fig. 2. Scroo Hill is a relatively exposed, rounded hill (height 248m, 813 feet) situated some 11 kilometres south of Lerwick. The Hill of Susetter is an elongated flat-topped hill (height 170m, 557 feet) with a steep western side and is situated some 20 kilometres north of Lerwick.

THE MONITORING EXPERIMENT

The choice of the equipment to be used in the experiment and the measurements to be taken were influenced by four basic factors: (1) The location of the sites: Not only is Shetland some 600 miles north of the base which controls the monitoring experiment but both hills are remote, thereby necessitating an independent power generation facility at each site and influencing the choice of the data storage medium. (2) The quality, frequency and type of meteorological data required. The monitoring is being carried out for several purposes - to provide a long-term estimate of the yield of wind turbines located on the hills; to record information for use at a later date by designers of wind turbines; and to record data for use in the time-step simulation model and the electrical quality model. These requirements conflict to some extent - in particular turbine designers are not only interested in the long term time series/mean pattern, but also extreme winds and turbulence patterns; the wind resource considerations must not only take account of time variation but also spatial variability, both horizontal and vertical; and the modelling required not only a long data time series but also fast data for the electrical quality model. (3) Finance - the money available was limited. Additionally any civil work carried out would include a "Shetland factor" due to the difficult access to the hill tops and the cost of shipping materials to the island from mainland Scotland. (4) The final factor, the climate, was in some ways the most important. Gusts of up to 49 m/s have been recorded at the Lerwick Met Office. Additionally Shetland receives relatively high amounts of rainfall (1318 mm in 1983) which combined with the high winds leads to problems of penetration of wind driven rain.

The project team decided to tackle the monitoring in the following way: (1) To establish a permanent monitoring station based on a fixed 45m mast at each site and to collect data from many different heights. (2) To purchase simultaneous and historic data from the two Meteorological stations on Shetland (see Fig.2) and (3) To carry out a wind survey across the summits of both hills using a portable 15m mast and 2 TALA kites. (The reason for

the wind survey is to check that the data recorded by the 45m mast is
representative of the wind regime above the hill tops).

The remainder of this paper will concentrate on the permanent monitoring
station. This consists of a guyed 45m lattice mast with instrument booms
every 5m from 10m upwards. Cup anemometers have been mounted on the booms
at 10m,15m,20m,25m,35m,40m and 45m, wind vanes at 15m and 45m, platinum
resistance thermometers at 10m,25m and 45m and 2 propeller anemometers
mounted at right angles to each other on the horizontal plane on the end of
the 30m boom. Fig. 3 shows the layout of the instruments. The data are
sampled with a data logger whose scan rate can be varied in 1 second steps
from a minimum of 1 second. The data are recorded on data cartridges by a
cartridge recorder and then posted back to RAL for decoding and analysis.

During the planning and installation of the monitoring station a number of
problems were encountered - these are now described together with the
solutions adopted, though it should be stressed that the solutions are
specific to the Shetland experiment and may not be universally applicable.

PRACTICAL PROBLEMS AND THEIR SOLUTION

Problem 1: At what rate should the 14 instruments be sampled?
If the sample rate is too slow the data collected will not be relevant to
all the applications, if it is too fast then frequent visits to the hill
tops would be needed to change the data cartridges. Our solution was
twofold, first to obtain a cartridge recorder which could write to ECMA46
type cartridges. The capacity of these cartridges is about 2.5MBytes which
is sufficient to hold data for around 14300 scans. The second part of the
solution was to decide to scan about once per minute (at this rate a
cartridge will last about 10 days) normally and to only scan at faster
rates such as 1/second when the research staff are on the island.

Problem 2: How could the risk of loss of data be minimised?
This problem was potentially very serious, given the time delay between the
recording of the data on Shetland and their subsequent decoding and analysis
at RAL. Our solution was to connect a radio transmitter to a spare port on
the data logger and to power the transmitter for three minutes every hour.
This results in 2 or 3 scans, which include the scan number, being
transmitted to the power station in Lerwick; thus enabling the power
station staff to visit the site just before the data cartridge becomes
full, and to be alerted about instrument failures.

Problem 3: How to supply power to the instruments, logger and recorder?
Both the sites are remote from power lines so clearly the power had to be
generated on site. The solution has been to install a bank of 4 12V sealed
lead acid batteries (capacity of each 110Ah), 2 50W Rutland Furlmatic Wind
chargers and a gas-fuelled generator and controller at each site.

Problem 4: Both the data logger and cartridge recorder have to be kept
within temperature limits (+5°C to +40°C) yet the power supply is
insufficient to heat the hut housing them. The solution has been to build
an insulated box, the walls of which contain removable slabs of expanded
polystyrene. The data logger and cartridge recorder reside in the
insulated box and so heat themselves. Using historic temperature records
we have prepared a guide showing which pieces of insulation should be in
place according to the month. As an additional check we have 2 max-min
thermometers, one of which hangs on the hut wall and one of which is placed
inside the insulated box. These are read and reset each time the data
cartridge is changed and the results used to adjust the level of

F1.3

insulation. The box has worked well, in winter, and also in summer when its thermal mass tends to damp the highest temperatures.

Problem 5: Lightning. Although both sites are not connected to the island supply system and are therefore electrically isolated much thought was given to the precautions needed against lightning. Lengths of scrap copper have been buried under the mast foundation block and the stay point blocks; these have been interconnected and fixed to the mast, the guy wires and the earth of the logging system. Additionally zener diodes have been wired into each signal line in an attempt to protect the logger and recorder from lightning damage.

Problem 6: Shetland's Climate. As previously mentioned, Shetland experiences high winds and a high rainfall thus creating the potential problem of penetration of water into instruments and cable joins. Initially we have avoided the need for junction boxes, which are particularly prone to water penetration, by having all instruments fitted with long lengths of cable sufficient to reach the hut which is situated 15m from the mast base. However, we recognise that it may be necessary to replace instruments later and are therefore researching into commercially available junction boxes and are also testing a design we have developed.

Problem 7: Wind Vanes. Several different types of wind vane are available - the more expensive versions are "self-referencing" i.e. they do not require to be aligned with respect to North. The project used a cheaper version which had linear output, apart from a 3 degree gap between 357° and 0°. Initially we tried to adjust the attachment of the wind vane to the mounting boom so that 0° as output by the vane matched True North. However, we found it more accurate to attach the vane to the boom and fix it at the correct height on the mast before determining the alignment of the vane. This method also provides a quick and simple check that the orientation of the vane does not change during the experiment, since the alignment check can be repeated whenever maintenance work is being carried out on the mast. Additionally a simple averaging routine has been written for the analysis software to determine the mean direction whilst taking account of the 360°/0° boundary.

Problem 8: Calibration of Anemometers. All the anemometers used have been calibrated in a wind tunnel at the Meteorological Office, as have a number of spare instruments. However, the problem remains of how to determine when an instrument needs replacing. The project has developed two tests to help answer this question, the first is a routine written into the analysis software which checks for anomolous wind speed readings. The second is a physical test which we plan to carry out at about 6 monthly intervals - we have designated one of our spare anemometers to be a standard. This instrument will be mounted on each of the instrument booms in turn for approximately 5 minutes and simultaneous measurements taken from the standard anemometer and the one being checked. Care will be taken to correct for the differing influence of tower shadow, if necessary, and then comparisons made between the 'standard' wind speed and the recorded wind speed.

Problem 9: Construction of a hill top monitoring station involves many different people and is dependant upon the weather. Initially it was planned to carry out the building materials to the sites before making and casting the concrete foundations. This proved to be impractical due to adverse weather conditions. A helicopter was employed to carry ready-mixed concrete and other items, such as the hut, to the hill top. This strategy proved to be cost effective and enabled the bulk of the construction work to be concentrated into a few days of good weather. The erection of the monitoring station itself involved many different teams of people - the

mast erectors, the research staff, the team installing the power supply and
the radio specialists. It was only by careful co-ordination that both
sites were installed with the minimum of delay.

Problem 10: Remoteness: The location of the sites has in many respects been
the biggest problem for the research staff. Amongst the steps we have
taken to minimise this problem have been to (1) test all the components
fully at RAL (2) have a full trial assembly at RAL (3) maintain a stock of
spare parts on Shetland (4) write detailed instructions about the assembly
of the system prior to leaving for Shetland (5) write extensive
documentation and to take many photographs (6) devise a method to examine
the data on site, for use whilst commissioning the apparatus (7) devise
hardware and software checks (8) write fault charts for the power station
staff to use in the event of a breakdown. Nevertheless, we are very aware
that the above steps are only effective as a result of the excellent
cooperation we receive from the power station staff who look after the
monitoring stations on a day to day basis.

RESULTS OBTAINED TO DATE

The above section has concentrated on the installation of the hardware.
The monitoring station at Scroo Hill has been operational since January
1985 and that at the Hill of Susetter from August 1985. We are therefore
able to present a few early results obtained from the analysis of the Scroo
Hill data.

Fig. 4 shows a histogram of the hourly mean values of the mean speed
recorded at 45m, the mean speed of the period being 9.60 m/s. Fig. 5
illustrates the distribution of wind directions recorded to date.

The analysis has also included the development of a method to resolve the
output of 2 Gill propeller anemometers mounted at right angles to each
other on the horizontal plane, to give wind speed and direction taking into
account their non cosine response.

ACKNOWLEDGMENTS

The authors gratefully acknowledge the help and advice given by other
project members (J. W. Twidell of the University of Strathclyde; N. Lipman,
R. Reynolds, R. Forbes and P. White of RAL; and W. Stevenson and G.
Anderson of the Hydro Board Headquarters). The Hydro Board were in
particular responsible for supervision of the civil engineering; provision
of the power supply and the radio apparatus; mast erection and the
day-to-day running of the apparatus. P. Fulton, J. Skelton and R. Sneddon
of Hydro Board; and J. Cousins and J. Watt of NSHEB Lerwick Power Station
are thanked for their help. The project also acknowledges the funding given
by the Science and Engineering Research Council under grant GR/C/82022 and
the North of Scotland Hydro Electric Board.

REFERENCES

Shetland Island Council (1985). Shetland in Statistics. Number 13 (1984).
ISBN 0-904562-20-4

F1.5

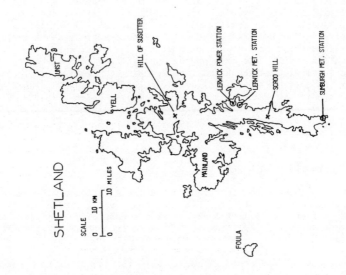

Fig 2 The Shetland Islands

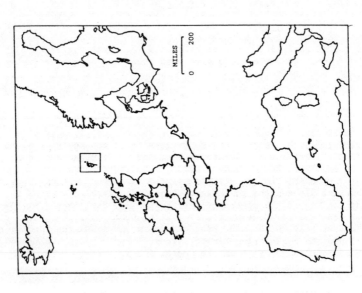

Fig 1 Location Map

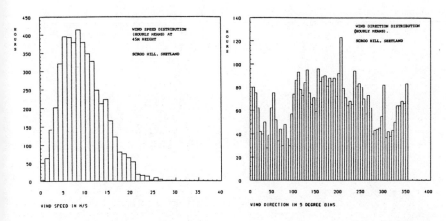

Fig 3   Fig 4

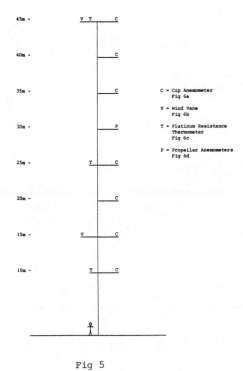

Fig 5

Fig 6a
Cup Anemometer

Fig 6b
Wind Vane

Fig 6c
Resistance Thermometer

Fig 6d
Propeller Anemometers

The Economics of Grid-connected Wind Turbines in Agriculture

D. I. Page*, E. A. Moughton** and R. Swansborough**

*Energy Technology Support Unit, Harwell, Didcot, UK
**ERA Technology Ltd, Leatherhead, UK

ABSTRACT

Calculations have been made of the import and export of electrical power on an hour-by-hour basis for eight types of farm (covering a range of sizes and types) when equipped with a WTG. These are based on the actual mean hourly power demands of the farms and the calculated mean hourly power generation by the WTG. Five types of WTG (10kW to 100kW) were considered separately for each farm and two wind regimes were used with mean annual wind speeds of $5.5 ms^{-1}$ and $7.0 ms^{-1}$.

The net charges to the consumer for each system before and after the installation of the WTG have been calculated using costs supplied by the WTG manufacturers and the tariffs for each of five Electricity Boards for the year 1984/85. A real rate of return on the capital investment has been obtained in each case.

KEYWORDS

Wind turbine; grid-connected; economics; electricity tariffs.

INTRODUCTION

Commercial developments in wind energy in recent years, particularly in Denmark and USA, suggest that small and medium-scale Wind Turbine Generators (WTGs) can be produced at prices which make them an economically viable alternative source of electrical energy, given a suitable financial environment. Following the Energy Act of 1983, there is now considerable interest in possible UK markets for these machines for connection to the National Grid and in the economic factors which would make their operation viable. Given economic attractiveness, large scale installation of such WTGs could result in them making a sizeable contribution (a few percent) to the UK national demand.

F2.1

Several broad-brush studies have been carried out to assess the likely
economic returns for close-coupled systems but all have lacked the
necessary detail to allow quantitative answers to be obtained. This paper
describes an in-depth study of grid-connected WTGs in one small sector of
the community --- farming, which was selected as the one most likely to
show economic promise.

ELECTRIC USAGE

Eight farm types were considered:

- small, medium and large poultry farms
- large dairy farms
- small, medium and large glasshouse 'farms'
- large pig farms

One case study was constructed to represent each farm type. Scenarios were
written to describe each farm in terms of type and size (eg crops and area
covered, stock and numbers and its energy-using activities). The pattern
of electrical power used on farms changes throughout the day and is also
highly seasonal; it was therefore necessary to build up profiles of
electrical power usage according to day and month and to time of day.

ELECTRICITY PRODUCTION

Five commercially available WTGs with rated output in the range 10kW to
100kW were considered. Costs were obtained from manufacturers or
estimated where this information was not available. The annual energy
production of each WTG was estimated using the power-wind speed
characteristic provided by the manufacturer combined with a wind speed/
duration curve to give a power duration curve. The power duration profile
is specific to the plant-site combination.

To enable the calculation of the nett electricity import-export of each
farm it was necessary to consider the WTG output hour-by-hour. To do this
wind data were obtained for an inland site with a mean annual value of
5.5ms^{-1} in the form of hourly means throughout the year. A second
wind profile corresponding to a mean annual windspeed of 7ms^{-1} was
derived so that two wind-time profiles were used for each case study.

TARRIFFS AND CHARGES

The fourteen Electricity Boards of England,Wales & Scotland have fourteen
different tariff structures and rates. Since much of the agriculture of
the UK is concentrated in four geographical regions - South Western
England, South Wales, North Western England and the South of Scotland, the
tariff structures used were those relating to these four regions with the
addition of the rest of Scotland.

The hour-by-hour energy output for each WTG was matched to the hourly
electrical energy usage on each farm and the tariffs applied to provide
annual electricity bills. Bills were also calculated for the case where a
windmill was not installed.

F2.2

ECONOMIC ASSESSMENT

The cost of providing the additional metering charges for the WTG operation
was added to the installed cost to give the total WTG installed cost; the
annual Running and Maintenance (R&M) costs were added to the annual
electricity bill to give the annual expenditure. This annual expenditure
was then compared with the electricity bill calculated assuming the WTG was
not installed, and the savings calculated. The profitability of the
investment was then calculated using a sinking fund rate of return (SFRR)
as the measure of profitability. A block diagram showing the flow of the
calculations is given in Table 1.

RESULTS

Rate of Return

One of the eight tables which summarise the results of the analysis is
shown in Table 2. This is the table for one farm type (large dairy farm),
divided into two parts for the different wind regimes. All figures are in
pounds sterling except the rate of return which is a percentage.

For each of the five WTGs the first section of the table gives the expected
rate of return on the installation under the tariffs and conditions of the
five Electricity Boards for the year 1984/85.   A negative electricity bill
indicates an excess of credit for units purchased by the Electricity Board
over charges for units supplied by the Board.

The rate of return is heavily influenced by the capital cost of the
installed machine, and both WTG-2 and WTG-4 are keenly priced for their
output. However, another important factor is the performance of the WTG,
particularly its output at the low wind speeds (which occur for much of the
time even in high wind speed areas), and the power-wind speed
characteristics show that WTG-2 and WTG-4 also have an advantage in this
respect. It must be stressed that the performance data are those supplied
by the manufacturers and not the result of independent tests.

In general the rates of return for WTGs installed for all farms in low wind
speed areas are negative; more economically viable rates of return of 5% or
greater are obtained only in the high speed areas.

Number of "Economic" Farms

The number of farms which can be classified under each case study was
obtained from the geographical distribution of farms by type and size
(provided by agricultural statistics) and a standard isovent map.

On the assumption that SWEB, South Wales EB and Norweb are typical of all
the Area Boards covering agricultural areas of England and Wales, Table 3
shows the numbers of farms in the UK where a rate of return of 5% or more
can be expected from the installation of a (suitable) WTG under the tariffs
operative in 1984-85. The total number of farms is less than 1300 and the
total annual energy production is 0.25 TWh compared to the national demand
of 275 TWH. Calculations to check the sensitivity of the analysis to the

input parameters showed that only a drastic improvement in tariffs (for example by making the unit purchase price equal the unit supply price) or a reduction in installed costs for the best WTG of 30% can increase the number of profitable farms by a factor of 10.

CONCLUSIONS AND DISCUSSION

It has been shown that in general the installation of small WTGs on a variety of farm types is either only marginally economic or uneconomic using the Electricity Board tariff structures and manufacturing costs for late 1984. However, there are of the order of 1,000 farms where installation of a suitable WTG would yield a rate of return on capital in excess of 5% and this number would increase sharply given more benevolent tariffs or a reduction in manufacturing costs. Currently the whole tariff philosophy of the Electricity Supply Industry is undergoing a radical change, this being the replacement of conventional Maximum Demand tariffs (MD) by the Seasonal-Time-of-Day (STOD) type; therefore the economic analysis reported here is a "snap-shot" of an everchanging situation. STOD tariffs may be of special significance due to the wide seasonal variation in the output of WTGs; it is possible that the new STOD tariffs will prove more favourable for private generation. Failing this, or a dramatic fall in manufacturing costs, it may be that as energy scenarios change, the ESI may adjust (or be encouraged to adjust) their tariffs to make private generation more economically attractive and hence increase the annual energy production from this source.

Given very favourable conditions, it appears that wide-spread installation of WTGs on farms could result in an annual energy production of about 1% of the national demand. Installation of a WTG in other building complexes (eg industry, hospitals) could result in this production increasing several-fold.

ACKNOWLEDGEMENTS

The results reported here are a summary of work carried out by ERA Technology Ltd under contract to the Department of Energy. It is intended to publish a full report of the work in due course.

TABLE 1   Flow Diagram for Calculating Rate of Return

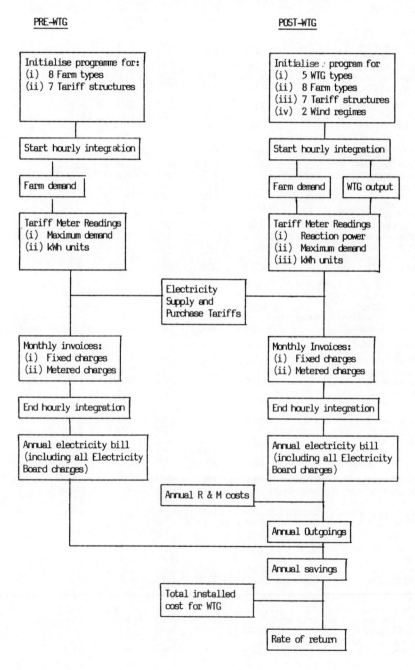

TABLE 2   Calculated Rates of Return for Large Dairy Farms

FARM TYPE 4 - LARGE DAIRY FARM (70,365 kWh pa)

ELEC. BOARD AMWS 5-6 m/s	WTG		ELEC BILL (pre-WTG)	ELEC BILL (WTG)	ANNUAL OUTGOINGS (WTG)	ANNUAL SAVINGS	RATE OF RETURN (%)
South Western	WTG 1	10kW	3285	3059	3377	-92	-3.43
South Wales	Inst. cost:		2975	2871	3189	-214	-3.98
North Western		£22,000	2957	2906	3224	-267	-4.22
North of Scotland	Running cost:		2739	2702	3020	-281	-4.28
South of Scotland		£318	3035	3781	4099	-1064	-7.81
South Western	WTG 2	40kW	3285	1370	1685	1600	2.09
South Wales	Inst. cost:		2975	1323	1638	1337	1.25
North Western		£31,000	2957	854	1169	1788	2.69
North of Scotland	Running cost:		2739	1447	1762	977	.10
South of Scotland		£315	3035	2555	2870	165	-2.49
South Western	WTG 3	55kW	3285	1269	1734	1551	.11
South Wales	Inst. cost:		2975	1197	1662	1313	-.37
North Western		£50,000	2957	888	1353	1604	.21
North of Scotland	Running cost:		2739	1277	1742	997	-1.01
South of Scotland		£465	3035	2010	2475	560	-1.89
South Western	WTG 4	60kW	3285	200	515	2770	3.87
South Wales	Inst. cost:		2975	205	520	2455	3.08
North Western		£40,000	2957	-485	-170	3127	4.75
North of Scotland	Running cost:		2739	703	1018	1721	1.26
South of Scotland		£315	3035	1834	2149	886	-.82
South Western	WTG 5	100kW	3285	-582	1446	1839	-1.59
South Wales	Inst. cost:		2975	-575	1453	1522	-1.83
North Western		£128,000	2957	-1420	608	2349	-1.19
North of Scotland	Running cost:		2739	136	2164	575	-2.57
South of Scotland		£2028	3035	287	2315	720	-2.46
AMWS 6-8 m/s							
South Western	WTG 1	10kW	3285	2086	3124	161	-2.30
South Wales	Inst. cost:		2975	2627	2945	30	-2.89
North Western		£22,000	2957	2634	2952	5	-3.00
North of Scotland	Running cost:		2739	2436	2754	-15	-3.09
South of Scotland		£318	3035	3531	3849	-814	-6.68
South Western	WTG 2	40kW	3285	22	337	2948	6.39
South Wales	Inst. cost:		2975	23	338	2637	5.40
North Western		£31,000	2957	-583	-268	3225	7.27
North of Scotland	Running cost:		2739	440	755	1984	3.31
South of Scotland		£315	3035	904	1219	1816	2.78
South Western	WTG 3	55kW	3285	-432	33	3252	3.53
South Wales	Inst. cost:		2975	-450	15	2960	2.94
North Western		£50,000	2957	-896	-431	3388	3.81
North of Scotland	Running cost:		2739	8	473	2266	1.55
South of Scotland		£465	3035	69	534	2501	2.02
South Western	WTG 4	60kW	3285	-1493	-1178	4463	8.07
South Wales	Inst. cost:		2975	-1444	-1129	4104	7.18
North Western		£40,000	2957	-2446	-2131	5088	9.63
North of Scotland	Running cost:		2739	-522	-207	2946	4.30
South of Scotland		£315	3035	-3051	10	3025	4.50
South Western	WTG 5	100kW	3285	-3304	-1276	4561	.53
South Wales	Inst. cost:		2975	-3260	-1232	4207	.26
North Western		£128,000	2957	-4288	-2260	5217	1.05
North of Scotland	Running cost:		2739	-1694	334	2405	-1.15
South of Scotland		£2028	3035	-2982	-954	3989	.09

F2.6

TABLE 3   Estimated Number of Farms Giving a Rate of More
          Than 5% and the Annual Energy Production

Farm Type	1984-85 Tariffs	
	No of Farms	Production (MWh)
Poultry (egg production)	71	14048
Poultry (broilers)	10	1979
Dairy	1125	222592
Glasshouse	73	14444
Pig	9	1781
All Farm Types	1288	254844

F2.7

Acoustic Noise from Wind Turbines: a Review

S. J. R. Powles

Sir Robert McAlpine and Sons Ltd., St. Albans Road,
Hemel Hempstead, Herts, UK

ABSTRACT

Acoustic noise from wind turbines could be a major source of environmental pollution. While the visual impact of large machines can be controlled by aesthetic design, particularly of the tower and nacelle, the noise of continuous operation is wanted by no one and could jeopardise planning applications. This paper reviews noise related problems experienced with wind turbines. Noise propagation and human annoyance depend on frequency and directionality. Noise generation is both mechanical (gearbox, generator) and aerodynamic (blade swish, tower shadow thump).

The physics of aerodynamic noise generation, directionality and frequency dependence is presented and discussed. Tower shadow noise seems to be a particular problem, with the low frequency noise generated as the blades pass near the tower propagating over considerable distances. Noise reduction and prediction techniques are discussed, as well as the legislation and regulations governing noise emission. A comprehensive bibliography is given.

KEYWORDS

Acoustic noise; wind turbines; environmental problems; siting; regulations.

INTRODUCTION

There are many environmental factors to be considered when siting a wind turbine : visual impact, noise generation, electromagnetic interference, bird life and risk to third parties. It is important that these issues are addressed if the introduction of wind turbines in the U.K. is not to be met with adverse public reaction. This is particularly true for rural and island communities, where the potential wind power may be reduced because of siting restrictions. Robson (1983) and Manning (1983) discuss these environmental problems.

Complaints about noise from wind turbines have been voiced in nearly all
countries developing this new energy source. Annoyance is subjective, but
noise can be measured and can be reduced by careful design. Most countries
have regulations covering noise emission and wind turbines must be
engineered to comply.

HUMAN SUSCEPTIBILITY

Noise can often be indicative of a problem with the structure of a wind
turbine. For example, a noisy gearbox would lead to further investigation
of its workings. Apart from this, however, the only reason to be concerned
with noise is because of human (and sometimes animal) annoyance.

The standard A-weighted sound pressure level definition (db(A)) was
introduced in the 1930s to weight sound amplitudes at different frequencies
to account for the variation of human frequency sensitivity, and this is now
universally used for measurements. Both physical response of the ear, and
subjective reaction to the sensation are incorporated into the weighting.
Background noise levels can vary considerably, from around 20 dB(A) in
isolated countryside, up to 70 dB(A) for a house on a busy road, (Allen
1982). A doubling in sound pressure level gives an increase of 6 dB.
Background noise will determine the nuisance of the impressed sound to the
community, particularly at night.

Higher frequencies are found to be the most subjectively annoying (Pearsons,
Bennett and Fidell 1979), but these attenuate most with distance through the
air and will only be heard near the turbine. The lower frequencies will
carry further, but are less noticeable to the population, apart from the
important case of infrasound (air pressure disturbances which occur at
frequencies below normal human detection limits) , which can be damaging to
both people and buildings (Kelley, Hemphill and McKenna 1981). Infrasound,
caused for example by tower shadow effects, can also lead to "secondary
sounds" (Keast and Potter 1980) such as rattling crockery and windows.

These frequencies propagate so far that complaints of resonances in
buildings several miles from the source have been received (Kelley 1981).
Horizontal axis downwind rotors are most likely to produce infrasound caused
by periodic chopping of the blades through the tower shadow wake. It seems
likely that a simple overall measurement of noise (e.g. dB(A)) will not
predict the audibility of wind turbines, particularly their low frequency
impulsive noise and broadband noise (Stephens and colleagues 1982).

MEASURING NOISE

Recommended practices for measurement of noise emission from wind turbines
have been minutely detailed in the recent expert study group report of the
International Energy Agency (Ljunggren, Gustafsson and Trenka 1984). This
document gives guidelines for noise measurement intended to facilitate
comparisons of data taken in different countries by different investigators.
A secondary goal of the regulations put forward is to provide an engineering
data base for analytical acoustic prediction techniques to be developed and
validated. However, the problem of acoustic acceptability levels for
regulations, and the "psycho-acoustic" aspect are not addressed. Necessary

measurement equipment is detailed for the determination of the A-weighted equivalent continuous sound pressure level, of third-octave band spectra, of narrow band spectra and of filtered instantaneous sound pressure. The requirements of International Electrotechnical Committee (IEC) Publications 225 (1966) and 651 (1979) will need to be met by the instrumentation. The preferred recording height is 5m above ground, which could cause problems with mounting heavy equipment, but this is an important parameter as ground proximity effects can influence measurement significantly.

Along with magnetic tape recordings of sound intensities, the following parameters will also need to be measured continuously: wind speed and direction at hub height, power output and RPM, also blade pitch angle and yaw angle if appropriate. Every half hour records should be taken of humidity, temperature, barometric pressure, turbulence and possible temperature inversion in the atmosphere.

Background noise levels are an important factor to consider when measuring the impact of a wind turbine on a community. Since wind turbines only operate in moderate to strong winds, evaluation of the masking effect of background noise introduces special problems not usually encountered in noise assessment. For every doubling of wind speed above 5 m/s, an increase in noise of about 12 dB(A) has been reported by Soderquist (1982a) and rain can have an effect of up to 15 dB(A).

As an alternative to making a series of measurements a few minutes long on several occasions, a recording system can be left at a site for many days recording data over a wide range of conditions. The CEGB use such a system, controlled by microprocessor, to collect samples for subsequent analysis (Nairne 1983).

MECHANISMS OF NOISE GENERATION

Noise can be generated both mechanically and aerodynamically, and each generation method will have a particular directionality and frequency characteristic.

Higher frequencies attentuate most with distance, absorption depending on temperature and relative humidity. Quantitative results and tables are given in Piercy and Embleton (1977). Other factors determining propagation distances are wind speed and temperature gradients locally, and interaction with the ground and other obstacles at site (Thomson 1982, Thomas and Roth 1981)

Obvious sources of mechanical noise are gearbox and generator, but vibrating structural panels and bearings can generate appreciable noise. Conventional soundproofing techniques can usually be applied to mechanical noise problems. Characterised by a fluctuating mass and applied force, mechanical noise is often monopole, and therefore an efficient converter of system energy to sound energy. A recent example was the WEG 25m Ilfracombe wind turbine, where a problem with the gearbox caused complaint. This also underlined the importance of local geography and habitation to siting. The Growian 3MW unit also had gearbox problems, necessitating isolation of its gearbox from the nacelle to reduce noise and allow it to operate at night (local regulations limit noise to less than 45 dB(A) at 250m distance).

The study of aerodynamic noise had been dubbed "aeroacoustics" (Goldstein 1974) or "sound generated aerodynamically" (Lighthill 1952). Noise is generated by tower shadow effects at very low frequency, and by aerodynamic turbulence and vortex effects over a broad frequency band. Dependent design parameters are RPM, power output, turbine diameter, number of blades, and blade chord (Engstrom and Gustafsson 1984).

Tower shadow noise led to worries about downwind HAWTs. This noise source depends on tower size and shape, blade passing distance from the tower, RPM and rotor diameter (Marcus, Marion and Harris 1983, Martinez, Widnall and Harris 1981, Vitenna (1982). Problems with the MOD-1 machine at Boone, N. Carolina, led to a speed reduction from 35 to 24 RPM to reduce noise, with a corresponding generator size reduction from 2MW to 1.5MW (Wells 1981). Tower shadow "thump" from WTS-4 can be heard up to 3 km downwind (Shepherd and Hubbard 1983), and from the Swedish Maglarp wind turbine up to 2 km away (Soderquist 1982b) where complaints have been received. Upwind rotors do not produce much marked impulsive noise (Spencer 1981, Viterna 1981).

Broad band noise ("swish") generation mechanisms can be compared with helicopter rotors (George and Chou 1983). High frequency vortex noise and rotational noise will both occur. The assumption that vortex noise is produced by random fluctuations of lift forces due to inflow turbulence over the blade leading edge, and convection of the turbulent boundary layer, implies a horizontal dipole directivity, maximum along the rotor axis, and zero in the plane of the rotor disc. It also implies a noise power level proportional to the sixth power of the blade tip speed, making this the overriding design parameter. Rotational noise is produced by periodic lift and drag forces acting on the blade, and volume displacement, resulting in radiation of a periodic pressure disturbance at harmonics of the blade passing frequency. Using 3 blades (lower RPM for same power) or a variable speed generator (low RPM in low winds when background noise is low) can reduce broadband noise. The dominant noise from the MOD-2 is centred around 1 kHz (Hubbard, Shepherd and Grosweld 1981), and this high frequency broadband noise is also the dominant upwind and crosswind signal from the WTS-4. Noise can also be caused by blade surface joints (e.g. at variable pitch blade tip attachment points) or imperfections.

                        NOISE PREDICTION AND REDUCTION

Since the noise generation problems encountered with MOD-1, research work has been undertaken on infrasound prediction techniques. Metzger and Klatte (1981) and Meijer (1984) discuss prediction tools. Near field broadband prediction models include 'Inflow Turbulence' and 'Trailing Edge Noise', but agreement between measured and predicted data is often disappointing, and far field analysis is more complicated. The ISVR, Southampton, has recently carried out a study for the Department of Energy, based on the WEG 20m Orkney turbine.

Vertical axis wind turbines may offer significant advantages for noise reduction, due to their low RPM operation with limited tower shadow. The ground level gearbox and generator enable efficient conventional sound proofing and reduce noise propagation distance. The CEGB will continue to monitor their Carmarthen Bay wind turbine site during the commissioning of the 25m variable geometry VAWT. In general, careful aerodynamic design, particularly of junctions and blade tips, together with reduced RPM operation, could reduce noise problems.

LEGISLATION AND REGULATIONS

Recommendations for noise measurement have been detailed above. In most countries there are legal provisions on noise generation limits, applicable to wind turbines. Noise is measured in dB(A), with the permitted level varying with time of day and location. The local authority, acting on a complaint, can require the wind turbine operator to limit noise level, and can restrict operation. Most regulations ignore background noise, particularly increasing background noise with wind speed, and take a fixed level. Regulations vary according to country, but average limits are between 35 and 70 dB(A) (depending on time of day) in a 6 m/s wind measured 0.5 to 1.5m above ground, 2-3m from the nearest reflecting surface. Acceptance criteria and standards are given by Stephens and Shepherd (1981) and Stephens, Shepherd and Grosveld (1981).

CONCLUSIONS

Noise from wind turbines could lead to siting restrictions and adverse public reaction, particularly in rural and island communities. Mechanical, tower shadow and broadband noise all contribute.

More research is necessary, particularly to try to predict and measure the noise generated by various blade designs, with the trailing edge and tip apparently the most important areas to study. Otherwise, only the use of multi-blade, low tip speed ratio rotors will effectively reduce the noise problem.

REFERENCES

Allen, J.E., (1982). Aerodynamics. Granada Publishing.
Engstron, S. and Gustafsson, A., (1984). General environmental aspects of large scale wind energy utilisation. National Energy Administration, Sweden.
George, A.R. and Chou, S.T., (1983). Comparison of broadband noise mechanisms, analyses and experiments on helicopters, propellors and wind turbines. AIAA-83-0691. Presented at 8th Aeroacoustic Conference, Atlanta, Georgia.
Goldstein, M.E., (1974). Aeroacoustics. U.S.A. NASA SP346.
Hubbard, H.H., Shepherd, K.P., and Grosveld, F.W., (1981). Sound measurements of the MOD-2 wind turbine generator. NASA CR 165752.
Kelley, N.D., Hemphill, R.R., and McKenna, H.E., (1981). A methodology for assessment of wind turbine noise generation. Presented at 5th Biennial Wind Energy Conference and Workshop, Washington D.C.
Kelley, N.D., (1981). Acoustic noise generation by the DOE/NASA MOD-1 wind turbine. Proceedings of 2nd NASA/DOE Wind Turbine Dynamics Workshop, Cleveland, Ohio.
Lighthill, M.J., (1952). On sound generated aerodynamically. Proc. Roy. Soc. A211, pp 564-87, and (1954) A222, pp 1-32
Ljunggren, S., Gustafsson, A., and Trenka, A.R. (1984). Expert Group Study on recommended practices for wind turbine testing and evaluation. 4, Acoustics measurement of noise emission from wind energy conversion systems. I.E.A. Programme for R & D on W.E.C.S.
Manning, P.T., (1983), The environmental impact of the use of large wind turbines. Wind Engineering 7, No. 1, pp. 1-11.

Marcus, E.N., Marion, M.A., and Harris, W.L., (1983). An experimental study of wind turbine noise from blade-tower wake interaction. AIAA-83-0691. Presented at 9th Aeroacoustic Conference, Atlanta, Georgia.

Martinez, R., Widnall, S.E., and Harris, W.L., (1981). Predictions of low frequency and impulsive sound radiation from horizontal axis wind turbines. Proceedings of 2nd NASA/DOE Wind Turbine Dynamics Workshop, Cleveland, Ohio.

Meijer, S., (1984). Wind turbine noise - prediction tools and design parameter dependence. Aeronautical Research Institute of Sweden.

Metzger, F.B., and Klatte, R.J., (1981). Status report on downwind horizontal axis wind turbine noise prediction. Proceedings of 2nd NASA/DOE Wind Turbine Dynamics Workshop, Cleveland, Ohio.

Nairne, P., (1983). The measurement of background noise levels at the aerogenerator site, Carmarthen Bay Power Station. CEGB SW Region report SSD/SW 82/N147.

Pearsons, K.S., Bennett, R.L., and Fidell, S.A., (1979). Initial study of the effects of transformer and transmission line noise on people. EPRI EA-1240.

Piercy, J.E., and Embleton, T.F.W., (1977). Review of noise propagation in the atmosphere. J. Accoust. Am 61, No. 6.

Robson, A., (1983). Environmental aspects of large scale wind power systems in the U.K. I.E.E. Proceedings 130, Pt. A, No. 9.

Shepherd, K.P., and Hubbard, H.H., (1983). Measurements and observations of noise from a 4.2 Megawatt (WTS-4) wind turbine generator. NASA CR 166124.

Soderqvist, S., (1982a). Swedish WTGs - the noise problem. Proc. 4th International Symposium on Wind Energy Systems. Stockholm, Sweden. B.H.R.A. (p.367).

Soderqvist, S., (1982b). A preliminary estimation of the expected noise levels from the Swedish WECS prototypes, Maglarp and Nadsudden. The Aeronautical Research Institute of Sweden. FFA TN 1982-01.

Spencer, R.H., (1981). Noise generation of upwind rotor wind turbine generators. Proceedings of 2nd NASA/DOE Wind Turbine Dynamics Workshop, Cleveland, Ohio.

Stephens, D.G., and Shepherd, K.P., (1981). Wind turbine acoustic standards. Proceedings of 2nd NASA/DOE Wind Turbine Dynamics Workshop, Cleveland, Ohio.

Stephens, D.G., Shepherd, K.P., and Grosveld, F.W., (1981). Establishment of noise acceptance criteria for wind turbines. 16th Intersoc. Energy Conv. Eng. Conference, Atlanta,

Stephens, D.G., Shepherd, K.P., Hubbard, H.H., and Grosveld, F.W., (1982). Guide to the evaluation of human exposure to noise from large wind turbines. NASA TM 83288,

Thomas, D.W., and Roth, S.D., (1981). Enhancement of far-field sound levels by refractive focusing. Proceedings of 2nd NASA/DOE Wind Turbine Dynamics Workshop, Cleveland, Ohio.

Thomson D.W. (1982). Noise propagation in the atmosphere's surface and planetary boundary layers. Internoise 82, San Francisco, U.S.A.

Viterna, L.A., (1981). Noise generation of upwind rotor wind turbine generators. Proceedings of 2nd NASA/DOE Wind Turbine Dynamics Workshop, Cleveland, Ohio.

Viterna, L.A., (1982). Method for predicting impulsive noise generated by wind turbine rotors. Inter-Noise Conference, San Francisco, U.S.A., 17-19 May.

Wells, R.J., (1981). GE MOD-1 noise study. Proceedings of 2nd NASA/DOE Wind Turbine Dynamics Workshop, Cleveland, Ohio.

IEC Publication 225. (1966). Octave, half-octave and third-octave band filters intended for the analysis of sound and vibration.

IEC Publication 651. (1979). Sound level meters.

The Establishment of the UK National Wind Turbine Centre

J. W. Twidell

Energy Studies Unit, University of Strathclyde, Glasgow,
UK

(notes from a presentation by G. Elliot)

SUMMARY

This paper describes the establishment of the UK Centre and test site for the development of small to medium size wind turbine. The role of the British Wind Energy Association (B.W.E.A), the Scottish Development Agency (S.D.A.) and the Department of Trade and Industry (D.T.I.) is summarised. Details of the Myres Hill test site are given, together with information about the National Wind Turbine Centre (NWTC).

KEYWORDS

WIND TURBINE TESTS DEVELOPMENT

1.   INTRODUCTION   Looking back over the last 3 years, it is now possible to describe the successful cooperation between a professional organisation (the BWEA), a national development organisation (the SDA) and a government ministry (the DTI). However in 1983 it was not possible to foresee that a campaign to lobby government could be so successful in so short a time.   This paper is written by one of the lobbyists who attempts to give a fair summary of the events.

The paper also includes details of the Centre taken, with acknowledgement, from papers published and circulated by the DTI.

At the 1985 E.R.I.C. Conference details of the Centre and grants available to manufacturers were given by Mr George Elliot, Manager of the NWTC.   A formal paper was not presented due to pressue of time, so this paper has been included in the Proceedings with his support.

2.   THE ROLE OF THE B.W.E.A.

From its inception in 1978 until 1983, the major emphasis of the British Wind Energy Association was to promote the case for large megawatt scale wind turbines.   This policy was reasonable in view of the Department of Energy's committment to power generation on a scale significant for the UK, and eventually helped the decision to fund the Wind Energy Group's 3MW aerogenerator project on Orkney.

208 J. W. Twidell

At the Annual Conference of the BWEA in Easter 1983 (Musgrove 1983), members voiced the need for more attention to be given to small machines to a capacity of about 100 kW. Consequently the Small Wind Turbines Group of the BWEA was established. This group drafted strategy paper which was effectively completed at the 1983 ERIC III Conference in Inverness (Twidell 1984), and incorporated in the BWEA booklet "British Wind Energy - A Development Strategy" in the UK" (BWEA 1984). Present at the BWEA and ERIC III Conferences were leading members of the Dept.of Energy associated with wind turbines, senior engineers from the Dept. of Trade and Industry and senior representatives of the Scottish Development Agency.

One major aspect of the lobby for small machines was the need for a UK test station similar to that at the Riso National Laboratory in Denmark and at Rocky Flats in the U.S.A. Mr Richard Morris of the SDA responded to the request by arranging for the Agency to fund the site works of a test site near Glasgow on Myres Hill. This site had been identified by Mr George Elliot of the National Engineering Laboratory, East Kilbride. The Laboratory is operated by the D.T.I. and consequently this Department allocated resources for the operation of the site, as described later.

Thus within a period of one year, what started as an expression of need by a majority of BWEA Conference members had materialised into a significant national development. The following detail of the Centre is taken from papers published or circulated from the National Engineering Laboratory.

3. THE NATIONAL WIND TURBINE CENTRE

(a) Introduction

Early in 1984 the National Engineering Laboratory supported by the Scottish Development Agency applied to the Department of Trade and Industry for support to set up a test centre for small wind turbines. Approval was given in May and work immediately commenced towards creating the facility.

The need for such a facility had already been identified by other countries, notably Denmark, Netherlands and USA for the support of their developing manufacturing industry. In UK however, development has been supported by the Department of Energy mainly investigating machines for grid-connection and for operation by the utilities. The need for wind turbines for overseas markets and for individual use in UK has been emerging and has received some stimulation by the 1983 Energy Act. In addition the recession in trade has led to an increase of the number of companies seeking diversification into new areas and away from their traditional product range, as well as new companies formed by brave entrepreneurs to tap what could be a major growth area.

The British Wind Energy Association has strongly supported the case for a national facility that would greatly enhance the infant industry and assist its growth. These were the primary reasons used to seek the support of the DTI which has been given for three years. After that time, if continued support can be justified, the manufacturing industry must have demonstrated

F442

significant sales and growth for the home and overseas markets.

(b) Objectives

The NWTC has several objectives, the principal one being to provide technical facilities and expertise in support of UK manufacturers. This can be achieved in several ways:

a   by providing independent accrediation of machine performance and quality as an aid to selling;

b   by providing engineering and technical expertise to improve the design and cost effectiveness of wind turbines and

c   by assuring high standards of materials and quality of manufacture compatible with high reliability

NEL intend to work in conjunction with other organisations that have expertise on wind turbines and related technology to develop engineering knowledge and to work towards the creation of standards covering quality, safety and operation.   Work on standards will recognise the needs of British manufacturers but compatibility with European, US and IEA standards will be sought.

(c) The Myres Hill test site

The Centre will encompass the whole range of mechanical engineering expertise already available at NEL especially in the fields of power transmission systems;   materials and structural testing;   control systems ; engineering and metrology.   In addition NEL intend to operate a test site for both short-term and long-term evaluation of industrial machines. The site is located 16 km south of Glasgow on a high moorland ridge, 340 m above sea level.

The site commands an uninterrupted open outlook on all sides, the land gradually sloping to the SW coast 50 km away.   No significant wind data have yet been collected for the site; however, data from East Kilbride to the NE (elevation 240 m) indicate an annual average in excess of 6 m/s, and winds in excess of 50 m/s are experienced on two or more occasions annually.

The site has been constructed in two phases. Phase I consists of three universal test pads capable of accepting machines of around 20 m diameter to the full survival wind load of 60 m/s.   Larger machines could be accepted, subject to wind loading calculations for individual designs. A further test pad has been constructed to the same design, but is located within the NEL grounds at East Kilbride.   The test site has a meteorological tower 50 m in height and anemometers spaced at various intervals in order to provide long-term records and wind profiles. Portable anemometer masts will be positioned at hub height immediately up-wind of each machine on test.   It is intended that the site will operate for periods while unattended.   Laboratory accommodation in the form of Portakabins is provided for personnel and equipment.

Phase II will proceed if the demand for the site and the growth in the industry warrants it. A virtually identical arrangement extending to the south is intended and the arrangements to lease the necessary land have been made.

Each wind turbine pad is supplied with a small instrumentation and data conditioning cabin which is connected by underground cable to the main laboratory accommodation. Measurements from the rotating parts of the machine are in digital form and the others are retained in analogue form. In the Portakabin all signals are muliplexed and encoded for transmission to the National Engineering Laboratory at East Kilbride. This 5 km line-of-sight transmission is via a UHF data telemetry link. All signals can be examined in analogue form on the test site.

The signals are received at the NEL Control Centre and are decoded prior to immediate analysis in a dual computer data system. The raw data are also put into store for further examination, if required, or in event of a computer malfunction. Status and alarm signals indicating a wind turbine malfunction are incorporated in the data stream.

A telemetry control link from NEL to the test site will allow data system checks, such as calibration, to be carried out remotely. The initial data system will be capable of handling 64 channels simultaneously over the UHF link; however, local processing on site is also possible. Data from each machine can be sampled at rates appropriate to the measured parameters but most measurements are sampled at 12 Hz. A channel for high frequency transients with a bandwidth of 200 Hz is also incorporated in the telemetry link.

Initially it is intended that machines with induction generators will be connected for supply to and from the UK grid. Autonomous generators will be connected to individual resistive load units appropriate to their output. Water pumping machines will be able to be accommodated with the provision of a 10 000 gall reservoir on the site. This reservoir would also be used for machines intended for direct heat production.

The site provides the basis for a wide range of future research and development activities, and additional facilities could be incorporated if the need arose. Discussions are taking place with other research organizations with regard to future R & D work, and it is possible that a basis for university research activity will be established in parallel with NEL's commercial interests. NEL looks forward to an active involvement by the manufacturing industry and centres of research in further wind turbine technology in the UK.

4.   CONCLUSION

Government ministries and organisations often have to bear strident criticism from lobbyists. In this particular case, it is a pleasure to acknowledge the rapidity with which the arguments of a professional organisation have been turned into the reality of a national test and development facility.

REFERENCES

1.   Musgrove, P (1983).   Proceedings of the Annual Conference of the British Wind Eenergy association, Cambridge University Press, Cambridge.

2.   Twidell, J W, Riddoch F, and Grainger W (1984). Proceedings of the Third conference "Energy for Rural & Island Communities", Pergamon Press, Oxford.

3.   British Wind Energy Association(1984) "British Wind Energy – A Development Strategy", obtainable from the BWEA, 4 Hamilton Place, London, published at the University of Strathclyde, Glasgow.

Power Matching Between a Small Wind Turbine Generator and Battery Storage (Patent Pending)

J. C. German

Northern Lighthouse Board, Edinburgh, UK

ABSTRACT

This paper refers to an electronic device intended to extract the maximum energy from a wind driven alternator over a wide speed range.

INTRODUCTION

This paper presents details of further development of the device described by German (1984).

It concerns the improved matching of a 50 Watt nominal wind driven permanent magnet alternator used for battery charging.

PRINCIPLE OF OPERATION AND BENCH TESTING

At a normal running speed, the EMF induced in the generator windings overcomes the battery voltage and delivers a normal charging current to the battery.

At higher speeds, to extract the increased power available, the generator voltage has to be allowed to rise, while the generator current is substantially constant. One way of achieving this is to insert an auto transformer between the generator and rectifier. The load can be switched between taps, while the generator remains connected to the same tap. By selecting the tap according to the speed, we can obtain a good match over a much wider speed range. We require a circuit which will sense the generator speed and switch in the appropriate tap automatically. Initial testing was done, however, with manual tap switching with the generator driven by a variable speed motor. The transformer used had 5 taps and for each tap, charging current was plotted against generator speed. This produced the family of curves illustrated in Fig. 1. The voltage ratios in this case were 1.0, 0.6, 0.4, 0.3, 0.2.

The frequencies at which tap changing should occur are determined from the points at which the curves intersect.

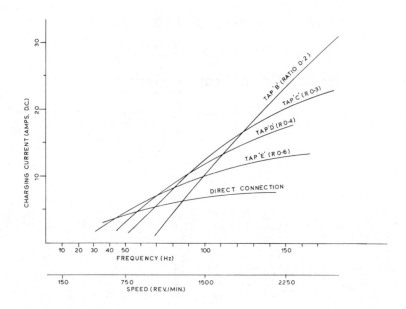

Fig. 1.

CONTROL CIRCUIT DEVELOPMENT

The Mark 1 and 2 circuits have been described in German (1984). This paper describes the Mark 3 circuit, which is an improvement on the earlier circuits as the number of IC's has been reduced from 22 to 4, and relays have been eliminated. Referring to Fig. 2, IC1 is a F/V converter. The output voltage represents the speed of the generator. This is applied to the non inverting inputs of the four comparators in IC2. The inverting inputs are set to the four voltages corresponding to the speeds at which it is desired that the taps change by means of four potentiometers. The DC voltage across the potentiometers is stabilised by a zener diode in IC1. The comparators are turned on in sequence as the speed increases, until they are all on at high speed. The ex-or gates in IC3 process the comparator outputs so that one of the five outputs is on at any time, representing the transformer tap called for. IC4 contains a buffer for each output. Each of the outputs 1 to 4 turns on a pair of opto isolated thyristors and an indicator LED. Output 5 turns on an LED only. When a pair of opto isolated thyristors are turned on, they turn on the associated pair of power thyristors, which form half of a bridge rectifier, so that the transformer tap connected to them is selected. The other half of the bridge rectifier consists of two of the diodes in an ordinary bridge rectifier. The other two diodes in the ordinary bridge are not forward biased and have no effect.

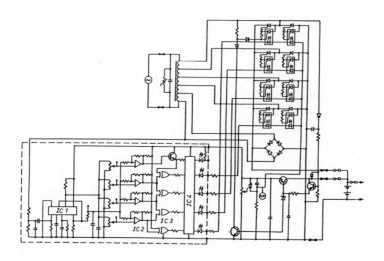

Fig. 2.

When tap 5 is called for, however, none of the thyristors is selected, and the ordinary bridge rectifier operates in the normal manner.

Two economy features included are a circuit to cut off the supply to the electronics when the generator is stopped and an oscillator which pulses the LED circuits at about 3kHz with a low duty cycle. The total consumption is 17mA with the generator running, zero when stopped. A circuit is also included to ensure that charging can take place even with a flat battery.

WIND TUNNEL TESTS

The complete controller incorporating the Mark 1 electronics with the transformer ratios previously mentioned was tested with the wind generator using the wind tunnel at the University of Newcastle. The controller performed satisfactorily and the results were as expected from previous tests with motor drive. There was very little change in speed caused by tap changing, the tip speed ratio being nearly constant.

F5.3

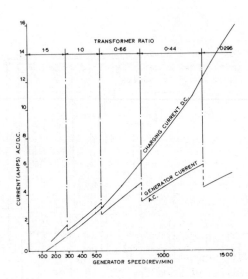

Fig. 3.

SITE TESTS

The above equipment was then installed at Cape Wrath and ran satisfactorily
for 3 weeks until damaged by lightning and high winds. The controller was
then converted to the Mark 2 electronics, which ran without trouble for 5
months. It was then converted to the Mark 3 electronics, and this was
installed in October 1984, and has been running without trouble since them.
A six channel recorder was added last February and the recordings confirm
correct operation.

LATEST DEVELOPMENT

A second controller with the Mark 3 electronics has been manufactured
recently with the electronics on a new PCB. A new transformer design is
used with the ratios 1.5, 1.0, 0.667, 0.444, 0.296.

The maximum step down ratio of 0.2 has been dropped as experience on site
shows this to be rarely used. The first ratio, however, is a step up,
intended to extract extra energy at low windspeeds.

F5.4

The performance of this controller using a motor driven generator is shown
in Fig. 3. The 'cut in' charging speed is reduced from a normal 200 rev/min
to 120 rev/min, the charge at 200 rev/min being about 0.5 amps.

It is intended that this will be a demonstration working model at the
Inverness conference, although it has not been wind tunnel or site tested.

OTHER APPLICATIONS

This device could be used with a wave powered generator, and with suitable
modifications with a three phase generator. If overcharging should be a
problem, a shunt regulator could be added.

CONCLUSIONS

The tests on the prototype controller have shown that a useful increase in
energy output can be obtained, of up to 400% instantaneous, and 50% average
over 3 weeks. Charging is improved at low as well as high windspeeds.
Generator reliability should be improved, as by maintaining a full load,
overspeeding is reduced.

ACKNOWLEDGEMENTS

Marlec Ltd, Oakham, for supplying the generators used.
Peterson Electronics, Forfar, for design and construction of PCB.
Cumbernauld Transformers Ltd, for supply of prototype transformer
Mr M Craig for construction of prototype

REFERENCE

German, J C (1984) "An Electronic controller to Maximise Efficiency of
                 Battery Charging from a Wind Generator" in Alternative
                 Energy Systems, Eds Mike West Et Al, Pergamon Press,
                 Oxford.

Optimizing the Performance of Wind-driven Alternators

L. H. Soderholm

USDA-ARS, Iowa State University, Ames, IA 50011, USA

ABSTRACT

Efficiency of a wind-energy conversion system (WECS) rotor is greatly dependent on the ratio of the rotational or tip speed to the wind speed driving it. Unless some provision is made for varying the rotor pitch or
speed, optimum tip-speed/wind-speed (TS/WS) ratios normally will not be
maintained over a wide range of wind speeds. Maximum efficiency, therefore, will not be obtained, and less than optimum power output will result.
Approaches for controlling the field of an alternator driven by a fixed-
pitch rotor are described that optimize the TS/WS ratio and thus enhance the
efficiency of a wind system.

Improvements in power output of wind systems from use of the field controls
have been significant, ranging from 10% to 80%. At wind speeds of 8-12 m/s,
power-output increases have ranged from approximately 1 to 5.5 kW for a
typical 10-kW machine, depending on the wind system characteristics.

KEYWORDS

Wind system efficiency; wind-driven alternators; wind energy.

INTRODUCTION

One of the major restrictions on the widespread use of wind energy systems
as an alternative energy source is the economic feasibility of their use.
Any factor such as reduction in cost or an improvement of the wind-system
efficiency enhances the practicality of such applications.

In the course of research directed towards the use of wind systems for agricultural applications, efforts have been made to improve the performance
of wind systems with fixed-pitch rotors that drive alternators. The efficiency or coefficient of performance (Cp) of a wind system using an alternator driven by a fixed-pitch rotor is dependent on the ratio of the tip
speed of the impeller to the wind speed driving it. Optimum tip-speed/
wind-speed (TS/WS) ratios for a fixed-pitch machine normally will not be

maintained over a desired range of wind speeds. Maximum efficiency, therefore, will not be obtained, and less than optimum power output will result.
By providing an optimizing control of the TS/WS ratio, an improvement of
the overall efficiency usually can be obtained.

Other methods that utilize the output energy of a wind system in an efficient manner can also result in performance improvement. One such approach
for heating and cooling of structures has been the use of line-commutated
inverters supplied by the rectified output of the alternator to produce
utility-compatible power. The energy output is enhanced by applying the
wind-generated power to a heat pump.

                    IMPROVING WIND-SYSTEM PERFORMANCE

A number of wind-energy conversion systems (WECS) use field-controlled alternators to supply AC electrical power to stand-alone resistance-heating
systems or for AC electricity to be rectified as an input for synchronous
inverters. For both applications, the power output is directly related to
the alternator's output voltage. If the TS/WS ratio of WECS can be held at
an optimum value over the normal range of wind speeds encountered, maximum
efficiency or Cp and best overall energy output generally will be obtained.

When WECS use a field-controlled alternator to supply a load that is proportional to the applied voltage, the load presented to the rotor and,
thus, its rotational speed may be varied by controlling the output voltage
of the alternator. Two approaches using this basic principle for performance improvement of fixed-pitch wind systems have been developed by USDA-ARS
personnel from wind research conducted in cooperation with the Department of
Energy.

Originally an analog circuit for field control was designed and patented.
Briefly, the basic control system consisted of a wind-speed sensor and a
frequency-to-analog converter providing inputs to a comparator circuit. The
comparator output was used to apply the proper field voltage to the alternator. A block diagram of the field control is shown in Fig. 1. Wind speed
was measured by an anemometer to provide an analog signal, proportional to
wind speed. The frequency-to-analog circuit sensed the alternator frequency
from the AC output and thus provided a measure of the tip speed by suitable
scaling of the analog signal from the relationship of the rotor diameter and
the gear ratio between the rotor shaft and the alternator.

Both the tip-speed and wind-speed signals were applied to a comparison circuit designed to provide an input to the field-voltage control proportional
in magnitude to the covarying wind speed and tip speed. The comparison
circuit output was applied to the field-voltage controller, thereby controlling the field of the alternator and, thus, its output voltage. As the
output voltage of the alternator increased or decreased, the load on the
impeller varied in accordance with the wind speed and extractable energy.
The alternator field voltage was, therefore, controlled in such a manner that
a predetermined optimum ratio of TS/WS was maintained. The approach worked
satisfactorily to produce increases of power and total output for specific
WECS of 10%-40% over a range of wind speeds from 4-7 m/s.

The circuit constants for the analog controller, however, needed to be tailored to each specific wind system. Simplification and cost reduction objectives have resulted in the design of a second version of the field control employing a microprocessor as the control element. In the first

                                                              F6.2

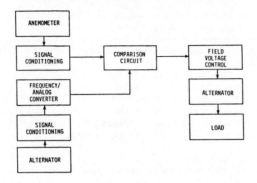

Fig. 1. Wind-system alternator field-control block diagram.

design, measurements of the wind speed and the rotational speed of the rotor
were required to maintain a TS/WS ratio at the optimum value by the field
control.   The second approach described achieves the objective of maintaining an optimum TS/WS ratio by maximizing the voltage across a load that has
a determinable characteristic.   Because power output is a function of output
voltage from the alternator, maximum power will be obtained if the voltage
output of the alternator is maintained at its maximum value.

A block diagram of the microprocessor control system is shown in Fig. 2.   For
control operation, periodic samples of output voltage are taken to determine
if incrementing or decrementing the field voltage will produce a higher
voltage than a previous sample taken before the field voltage change.   If
incrementing the field voltage produces an increase in output voltage, the
field voltage is again incremented, and the output voltage checked to see if
a further increase in output voltage is obtained.   Further increments of
field voltage are made until a decrease in output voltage is obtained, which
would indicate that the rotational speed had been reduced below the optimum
value or that the wind speed had dropped and that the load presented to the
wind system should be reduced.

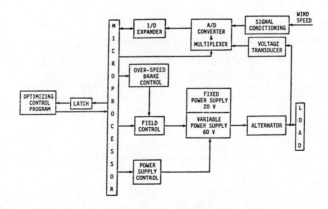

Fig. 2. Block diagram of the microprocessor field control.

Although a wind-speed input is indicated in the block diagram of the control
in Fig. 2, no measurement of wind speed is actually necessary for optimizing
the output voltage. Wind-speed measurement, however, speeds up the decision
process and provides for additional useful control functions such as brake
application for high wind speeds. The basic implementation of control is as
follows: 1) Wind speed is measured, and an approximate alternator field
voltage applied as determined from a lookup table contained in memory; 2)
alternator output voltage is sampled and verified that it is within specified
limits, or shutdown is initiated; 3) if the output voltage is within limits,
the field voltage is incremented and a determination made if a higher output
voltage results; 4) the field voltage is incremented again until a decrease
in output voltage occurs; 5) when a higher output voltage does not result
from incrementing, the procedure is reversed, and the field voltage is decre-
mented to determine if an increase in output can be achieved; 6) cycling
continues between incrementing and decrementing the field voltage to keep
the output voltage at a maximum value; and 7) a software hysteresis loop is
used to prevent unnecessary variations of field voltage due to small output
voltage changes.

CONTROL SYSTEM PROTECTION

Although concern has been expressed about possible damage to the control
circuit from lightning, no problems have been encountered. Several precautions were taken that possibly have helped to avoid damage. First, the
steel tower was grounded at each leg with a 10-ft ground rod. Second, alternator output and field wires were brought down inside the tower leg and
provided with thyrector surge suppression at the generator. And, third,
thyrector surge protection was provided at the AC input to the control circuit. Although little can be done about direct lightning strikes, these
measures have been effective in preventing control malfunctions. It is recommended that similar methods be used for all wind system installations where
potential damage from lightning exists.

RESULTS

Typical results for the use of the microprocessor field control on the output
performance of wind systems are shown in Figs. 3 and 4.

As shown by the data presented in Fig.3,   the increase in power output obtained by the use of the microprocessor control over that obtained by the
use of a fixed field of 35 volts ranged from 57% at 10 m/s to 137% at 5 m/s.
The increase in power output was approximately 0.76 kW at a wind speed of
5 m/s, 1.55 kW at 7.5 m/s, and 3.7 kW at 10 m/s. Performance improvement
of power output was greatest at the higher wind speeds when the fixed field
did not provide sufficient loading of the wind system and the rotor ran at
a higher than optimum tip-speed/wind-speed ratio. At wind speeds above approximately 11 m/s, the blades were feathered by a mechanical overspeed
system that limited the output for both systems.

A comparison of the performance obtained with the microprocessor control
with that obtained with the alternator in the configuration normally used
for input to a synchronous inverter is shown in Fig. 4.

The increase in output power by use of the microprocessor control over that
of the normal synchronous inverter output ranged from 47% at 5 m/s to 29% at
9.5 m/s. Although the controller did not provide the same degree of

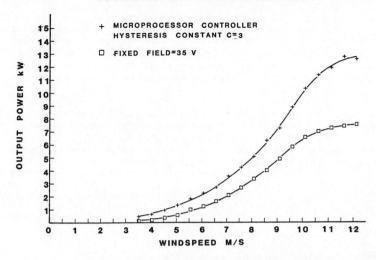

Fig. 3.   Alternator power output of a fixed-pitch wind system
          using a fixed field, as compared with output using a
          microprocessor-controlled field.

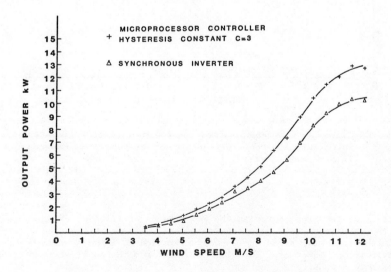

Fig. 4.   Alternator power output of a fixed-pitch wind system used as
          input to a line-commutated synchronous inverter for the
          normal field control function as compared with the use of
          the microprocessor-controlled field.

F6.5

performance improvement as it did when compared with the fixed field, the
improvement was significant. The synchronous inverter used for comparison
had a control circuit that automatically adjusted the field voltage in response to generator speed and power output as a control method to enhance
performance. Performance improvement would have been even more substantial
if the comparison had been made with a synchronous inverter without a field-
control function.

The value chosen for the voltage hysteresis constant "C" for the microprocessor control has an effect on the ability of the control to maximize
output, particularly at the higher wind speeds, as shown in Fig. 5. At
10 m/s, for instance, a hysteresis constant of three produced approximately
14% more output than a hysteresis constant of two.

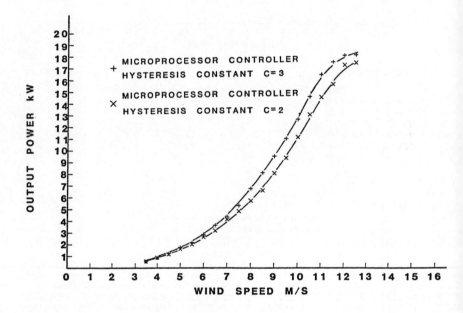

Fig. 5. Comparison of power output for the microprocessor control
         in relation to the hysteresis constant "C" for a Jacobs
         17.5-kW wind system.

SUMMARY

Two control systems for improving the performance of fixed-pitch wind systems with alternators have been developed for applications that use alternator output to supply resistance loads or provide AC input to line-commutated synchronous inverters. Results obtained from the application of these
controls have demonstrated substantial performance improvement of the wind
systems and indicate that the use of such control systems can enhance economic feasibility of the WECS. A USDA patent has been granted for the comparator system, and a USDA patent is pending for the microprocessor system,
thus making the developed principles available for all wind systems.

Educational Aspects of Wind-turbine Generation Using the Rutland Windcharger

A. Wheldon* and J. W. Twidell**

*Physics Department, University of Swaziland, Kwaluseni,
Swaziland
**Energy Studies Unit, University of Strathclyde,
Glasgow, UK

ABSTRACT

We have used this small commercial aerogenerator to investigate the basic
physical principles of wind-power generation, using only elementary measurement equipment available to schools.   The experiments include bench tests
which are safe and easily performed, and tests with the turbine mounted on a
car which could be performed under supervision.

KEYWORDS

Wind; electricity; educational; laboratory.

INTRODUCTION

Topics in renewable energy should be included in science and engineering
curricula at school and tertiary level, but a major problem is the provision
of suitable experimental work.   Difficulties include, (i) equipment is either
expensive or on such a small scale that it seems like a toy; (ii) the most
widespread sources of energy (solar and wind) are unpredictable for timetabled
programmes, and (iii) the experiments must be safe and easily performed.

The Rutland Windcharger (Marlec, 1985) is a commercial aerogenerator of 50 W
nominal capacity, designed for battery charging.   It is marketed internationally and costs about £200 (+ VAT).   We find it excellent for teaching
at a level and budget for practicals in schools and elementary tertiary
courses.   We have deliberately used the minimum of equipment, yet have been
able to derive experimentally many of the standard characteristics of windpower generation (Golding, 1976; BWEA, 1984; Twidell and Weir, 1985).

THE RUTLAND WINDCHARGER AND OUR ADAPTATIONS

Figure 1 is a photograph of the Windcharger, mounted for road tests, and
Fig. 2 is a circuit diagram.   It is robustly made, with a plastic-cased

Fig. 1.   The Rutland
          Windcharger, mounted
          for road tests.

permanent-magnet generator that rotates on a fixed shaft and forms the hub for
fixing the blades.   Six plastic blades are normally used but are easily removed, and symmetrical arrangements of 6, 3 or 2 blades can therefore be obtained.   Single-bladed operation with a counterweight is also possible.

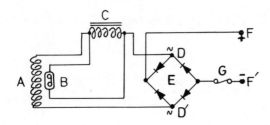

A. Stator windings
B. Protection thermostat
C. Control choke
D. a.c. output
E. Full-wave rectifier
F. d.c. output
G. Fuse

Fig. 2.   Basic circuit diagram of the Windcharger

The 'nacelle' contains the a.c. generator output via slip rings (needed when
the whole nacelle rotates in the wind), and a full-wave rectifier giving un-
smoothed d.c.   We made connections directly to the a.c. output for our
experiments, and therefore moved the fuse to the nacelle.

The hub presents a 25 cm diameter flat surface to the wind which cannot be
beneficial aerodynamically, but is very useful for bench experiments!   We
fixed a 21 cm diameter pulley to the front of the hub and drove it with a belt
drive from a variable-speed d.c. motor.   The blades were removed and the
nacelle mounted on a short length of 2-inch diameter pipe, clamped in a laboratory vice.   Safety guards should be fitted around the belt drive for
student use.   (Note that suitable motors are available from old washing
machines.)

A range of wind speeds is needed to study experimentally the characteristics
of the complete aerogenerator.   We did not have access to a wind tunnel, and
it is highly unlikely that elementary teaching laboratories would either.

We therefore mounted the Windcharger on an A-frame attached above the front bumpers of a car (Peugeot 404, with convenient jack-points and roof-rack to fix the frame), and provided a range of wind speeds by driving the car at different speeds along a straight and reasonably flat road.   The A-frame was held forward by a horizontal strut to the roof-rack, and this arrangement would allow the thrust on the turbine to be measured.   The output cables led to the rear passenger seat.

BENCH EXPERIMENTS AND ANALYSIS

A preliminary experiment related motor torque (measured approximately using a spring balance and a water-cooled rope round the pulley) to frequency (measured using a hand-held tachometer) and the current drawn by the motor.

We then drove the generator from the motor and measured at output the a.c. voltage (V) across a variable resistor; the current (I), and the generator frequency (f_g) using an oscilloscope (a frequency meter could be used but does not show the actual waveform).   We also measured the frequencies of the turbine shaft (f_t) and the motor using the tachometer, and the current drawn by the motor.   Measurements were made for resistive loads between 2 and 40 ohm and also on open circuit, for turbine frequencies up to about 15 Hz.   With the generator stationary, the resistance between the a.c. output terminals was measured as about 6 ohm, using an ohmmeter.

Generator Characteristics and Maximum Power Production

Load resistance (R = V/I) and power dissipated in the load (P = VI) were calculated, taking into account the small, but measurable, resistance of the connecting leads.   (Students should be made aware of power losses in leads.) Load resistance for maximum power increased slightly with frequency, from about 6.5 to 10 ohm (mean 8.0 $\pm$ 1.0 ohm).   Maximum power therefore occurs at loads close to the measured resistance of the windings.   Figure 3 shows that both the open-circuit voltage and the voltage for maximum power increased linearly with frequency: the numerical values agree well with the manufacturer's test results, in the frequency range which we were able to study (Marlec, 1985).   The ratio f_g/f_t is the number of magnetic pole pairs in the generator, which we measured as four to the nearest integer.

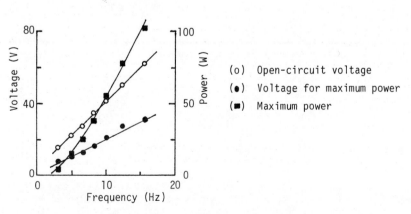

Fig. 3. Voltage and maximum power versus turbine frequency

Generator Efficiency at Maximum Power

The generator efficiency (η_g) is the ratio of the maximum power output
(Fig. 3) to the shaft input power from the motor, which is the product of
motor frequency and torque.   The calculated efficiency showed a small increase with turbine frequency and stabilised at about 60 ± 15 % between 10 and
15 Hz.   The approximate calibration gives a large experimental error, but the
results suggest that the generator efficiency must be close to its ideal
theoretical value of 50 % at maximum power.

Battery Charging

The Windcharger is intended for battery charging, but we recommend that this
is not studied in initial laboratory experiments, because the complex and
changing loads presented by rechargable batteries may distract from an understanding of the aerogenerator characteristics.   However, the principles
involved can be appreciated from our a.c. measurements.

A terminal voltage of about 13 V d.c. is needed to charge a nominal 12 V
battery, and 26 V is needed to charge two batteries in series.   The output
current at these voltages is plotted as a function of turbine frequency in
Fig. 4.   The charging efficiency η_c (which is the ratio of the power at the
charging voltage to the power into an optimum load) is also plotted.   It is
clear that the turbine frequency for maximum η_c increases as the charging
voltage increases.   At frequencies above 10 Hz it is more efficient to charge
two batteries in series than a single battery.   Further student discussion
about battery charging requires an understanding of tip-speed ratio and wind-
speed probability, as well as load matching (see German, 1984).

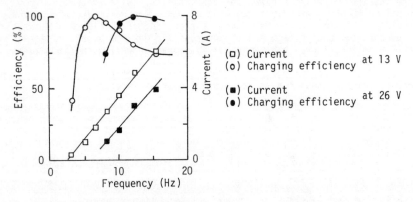

Fig. 4.   Current and charging efficiency versus turbine
          frequency, as for battery charging.

CAR TESTS AND ANALYSIS

The car tests were surprisingly easy and would add excitement to students'
normal laboratory experience.   Distance markers were established every 50 m
and were used both for the lower car-speed tests and to calibrate the speed-
ometer.   The driver measured the car speed, and the passenger measured the

voltage across a rheostat, previously set close to the maximum power resistance. For each car speed, we obtained an average of 5 voltage measurements driving in each direction along the road, at speeds between 4 and 25 m/s (8 and 50 mph). A full set of measurements was taken with the 6-bladed Windcharger, then repeated using symmetrical configurations of 3 and 2 blades.

The kinetic power in the wind of speed u incident on the turbine of disc area A is $p_w = 0.5 \rho A u^3$, where the air density ρ was calculated as 1.1 kg/m^3 for conditions in Swaziland at the time. We assumed that the wind speed at the turbine is equal to the speed of the car, although this is approximate due to the interference of natural wind and the pattern of circulation round the front of the car. (Direct measurement of wind speed would have been a distinct improvement, and could have been done using a hot-wire anemometer or Pitot tube fixed about 50 cm ahead of the turbine.) The fraction of kinetic power extracted as shaft power by the turbine is C_p, the power coefficient. The fraction reaching the load is $\eta_l = C_p \times \eta_g$, where η_g is the generator efficiency, defined previously. The output power and η_l are plotted as a function of car speed for 6-, 3- and 2-bladed operation in Fig. 5. The form of the curves shows the student that, (i) the output power increases rapidly with increasing wind speed, but the efficiency of conversion reaches a maximum and then falls, and (ii) as the number of blades is reduced, a higher wind speed is needed for maximum efficiency. Note that the output power increased continuously to about 500 W. The Windcharger has a thermal switch to add a choke in circuit if the generator overheats, and this would reduce the current. We clearly did not reach this state.

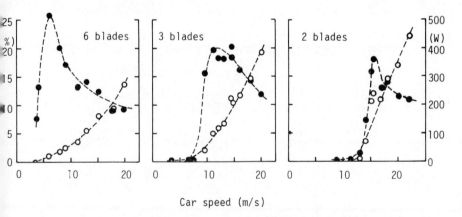

Car speed (m/s)

Fig. 5. Power output (o) and η_l ($\bullet$) versus car speed

Blade tip-speed is $2\pi r f_t$, where the blade radius r = 0.455 m. The turbine frequency was calculated from the voltage using Fig. 3 as a calibration. (We extrapolated to the maximum measured voltage of 67 V, because of the close agreement between our results and the manufacturer's tests.) Tip-speed ratio (λ) was calculated as the ratio of the average tip-speed (from the two directions of travel) divided by car speed. Figure 6 shows η_l plotted as a function of λ. The efficiency of the 6-bladed turbine shows a maximum at a tip-speed ratio of about 3. The results for the 3- and 2-bladed configurations are less conclusive, but indicate maximum efficiency at higher λ, as predicted by theory. It seems that the blades do not have optimum design for these configurations, because the tip-speed ratio for maximum power is not exceeded. Other blades could be made and tested.

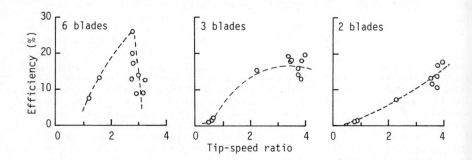

Fig. 6. η_1 versus tip-speed ratio

EDUCATIONAL ASPECTS AND CONCLUSIONS

A wide range of physical concepts is involved in these experiments, some with
general application in science education, and some specific to wind-power
generation.   Three areas could be developed.

Basic Concepts of Electrical Generation and Performance Matching

(a) The simple permanent-magnet generator demonstrates Faraday's law, that the
induced voltage is proportional to the rate of change of magnetic flux (Fig. 3)

(b) Bench tests demonstrate a.c. generation and rectification for d.c. use.

(c) The optimum load measured in bench tests is shown to be similar to the
internal resistance of the generator windings, measured directly.   For elemen
tary Physics classes, this can be demonstrated by power measurements at one
frequency.   For more advanced students there could be further consideration o
temperature and reactive effects, and of the meaning of 'efficiency'.

(d) Further study could include battery-charging, for which the Windcharger
is intended.

Power Conversion in an Aerogenerator

The theory of the experiments is based on the fundamentals of energy and mom-
entum conservation.   Students find it instructive to apply these principles
to wind-turbine performance.   The practical experience reinforces several
important concepts, namely (i) power in the wind varies as the cube of wind-
speed; (ii) useful power is obtained after a series of energy transformations,
and (iii) various efficiencies can be defined, and depend on load, turbine
frequency and tip-speed ratio.   It is of benefit for students to work with a
commercial machine, because they can easily extrapolate their experience to
larger-scale wind generation.

The experiments could be extended to include the measurement of the actual
air speed onto the turbine, and also the thrust on the turbine.

Instrumentation, Experimental Technique and Data Collection

We used only simple instrumentation, but the results permit analysis on a variety of levels.

The bench experiments gave very 'clean' results, but there was considerable scatter in the results from the car tests.   This can be used to stress the importance of repeat measurements in experimentation (an idea which is some-times neglected in Physics, where elementary experiments are often used for 'one-off' results).   For more advanced students, techniques for combining data could be considered in more detail, using the Windcharger results as an example.

An educational area which is expanding rapidly is the use of microelectronics for measurement and control.   Our experiments involved collecting many numbers and performing repetitive calculations with them: these are activities which could be used to demonstrate the application of data-logging and computer-handling of data.   Microprocessor control is essential for the commercial development of renewable-energy systems, and this could be demonstrated on a laboratory scale using the Windcharger.   One simple application would be a switching system to change the number of batteries on charge as the frequency of the turbine changed.

In conclusion, we find the Rutland Windcharger a most useful addition to an elementary teaching laboratory, which provides the opportunity to teach the fundamentals of wind-power generation within present curricula. A detailed study guide for pupils is being prepared based on this work. (Wheldon A., and Twidell, J.W.).
          ACKNOWLEDGEMENT

Our thanks to Professor Alan Ward for his assistance, and in particular for the loan and adaptation of his car.

          REFERENCES

British Wind Energy Association (1984). Wind Power in the 80s. N. Lipman, P. G. Musgrove and J. H. Pontin (Eds.) Peter Peregrinus, Stevenage.

German, J. C. (1984). An electronic controller to maximise efficiency of battery charging from a wind generator. In M. West (Ed.) Alternative Energy Systems, Pergamon Press, Oxford.

Golding, E. W. (1976). The Generation of Electricity by Wind Power. E. and F. N. Spon, London.

Marlec (1985). The Rutland Windcharger, manufactured by Marlec Ltd., Unit 5 Pillings Road Industrial Estate, Oakham, Rutland LE15 6QF, U.K.

Twidell, J. W. and Weir, A. D. (published 1985   ). Renewable Energy Resources,   E. and F. N. Spon, London.

Wheldon, A. and Twidell, J.W. "Wind turbine generator experiments". Booklet to be produced by the Energy Studies Unit, University of Strathclyde, Glasgow G1 1XQ.

TOPIC G

Wave Power

Wave Powered Desalination

S. H. Salter

Department of Mechanical Engineering, University of
Edinburgh, UK

ABSTRACT

Power from sea waves can be used directly for the production of fresh water,
thus avoiding some of the costs of intermediate electrical stages.

KEYWORDS

Wave energy; desalination; vapour compression; reverse osmosis; duck; heat
exchanger; mesh plastic; heat pump.

INTRODUCTION

Since 1973 a research group at Edinburgh University has been studying
mechanisms for generating electricity from sea waves with particular emphasis
on a device known as the duck. In accordance with the requirements of the
UK Department of Energy, the work was aimed at a large (2 GW) installation
off the Hebrides. Our interpretation (Salter, 1985) of the official
assessment methods produced a cost estimate of a little under 4 pence per
kilowatt hour for North Atlantic conditions. The possibility of a figure
below 5 pence' has, with reluctance, been accepted in an official report
(Davies, 1985). However, the UK has no present officially perceived need
of 2 GW of renewable energy and so we are turning our attention to smaller
installations and an alternative product.

The alternative product is fresh water. In many parts of the world –
particularly in island communities – oil is burnt to provide energy for
desalination so that a renewable energy source might be regarded as an oil
generator. Indeed as many water-deprived sites are in otherwise very
attractive places, the process can also be regarded as a land generator.

There are three well-proven techniques for desalination. They are multi-stage
flash, which uses large amounts of low-grade heat; reverse osmosis, which
requires high pressures at low flow rates; and vapour compression, which
uses high volume flow at very low pressure. Excellent descriptions of all
three techniques can be found in Porteous (1983) and the Proceedings of the
International Desalination Association conferences.

Both reverse osmosis and vapour compression are sometimes criticised for
their need for energy in the form of mechanical work. If this is provided
by electricity generated from fossil fuel, then the apparently attractive
overall energy efficiency is reduced by a factor of 3 or 4. However if this
mechanical work were to be derived directly from a renewable energy source
the objection would vanish. Furthermore, water is easily stored so that
the costs associated with irregular electrical output are removed.

THE DESALINATING VAPOUR-COMPRESSING DUCK

Figure 1 shows the simplified section of a falling film heat-exchanger which
in practice would consist of very many more plates. Hot brine is fed into
the unit through manifolds **A** and flows down the dashed path of the heat-
transfer surfaces. If a pump **B** were to reduce the pressure on the brine side
of the circuit, a part of the brine would evaporate. The pump would return
the vapour at a higher pressure and temperature to the upper side of the
heat exchanger. At this higher pressure it would condense on the slightly
cooler surfaces and thereby provide the latent heat for further evaporation
of brine. Condensate would fall and be removed from channels **C**. Excess brine
would be collected from the sump and returned to the upper manifolds **A**.

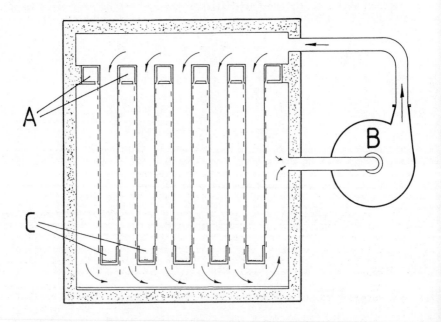

Fig. 1. The basic principle of vapour compression

In conventional equipment the vapour pumping is done by a high-speed
centrifugal compressor. The density of steam is low and the necessary
pressure rise - about a tenth of an atmosphere - is rather large for
centrifugal machines and so they have to run extremely fast. Their efficienc
is usually 70-80%. Problems can sometimes arise if water droplets hit rapidl
moving blades.

Much of this unpleasantness is removed if the pumping is done by duck motion.
Figure 2 shows a hollow duck about half full of water which will act both
as an inertial reference and as a double-acting piston. A heat-exchanger
is connected to the inner hull through passages containing non-return valves
D, E, F & G, drawn here in their symbolic electrical form. Suppose that
a wave lifts the beak of the duck so as to rotate it anti-clockwise. The
steam volume in the right half of the hull **H** will expand, drawing more vapour
from the lower section of the heat exchanger through valve **E.** Meanwhile
the volume of the left side **J** will reduce, thereby pumping vapour to the
top of the exchanger through valve **G.**

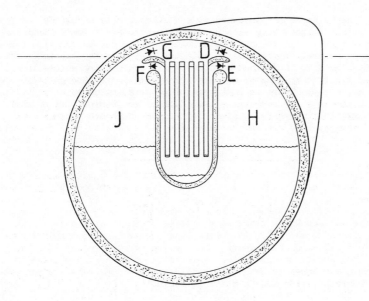

Fig. 2.   The hull of the duck provides a cheap
vapour compression pump.

The entire inner volume of the duck acts like an extremely large pump. The
low velocities point to high efficiency. There is no piston wear or
requirement for accurate machining. Given that the cost of the hull can
be charged to another account, the pump will be much cheaper than that of
a comparable turbo-compressor. However, we must pay very careful attention
to the design of the valves. They must operate several million times a year
with a very low pressure difference but not restrict the flow of vapour when
they are supposed to be open or leak when they are supposed to be shut.

The boiling temperature of the brine is slightly higher than the condensing
temperature of the pure vapour. For a concentration double that of sea water
this rise is 0.92C°. However, an additional larger temperature difference
will exist across the heat-transfer surface. This has to pass the full latent
heat requirement (2.26 MJ/kg at 100°C). If we know the surface area and
its heat-transfer coefficient we can calculate the corresponding temperature
difference. The temperature of saturated steam is completely defined by
its pressure and vice versa. We can get from one to another by looking up
steam tables. Hence we can obtain the pressure needed for any rate of water

G1.3

production. This pressure will be felt as a torque on the duck which tends
to oppose the motion given to it by the waves. Apart from a small dead space
in the middle of the operating band, which is caused by the elevation of
boiling point, the duck will feel a torque which is in proportion to its
velocity. This is nearly the perfect power take-off for a duck excited at
its natural frequency. We might extend the frequency range by using active
rather than passive valves and changing the time of operation. For best
efficiency, the power take-off damping coefficient should be made equal to
what the hydrodynamicist describes as the 'radiation damping' of the duck
shape. This is the ratio of torque to velocity that would be felt if the
duck were driven as a wave maker in calm water.

The size of the bulge on the waveward side of the duck is a powerful design
parameter. Small bulges produce large angular motions but low excitation
torque. We set the bulge to suit the damping coefficient to suit the pressure
to suit the temperature difference to suit the heat-transfer characteristics
at the heat exchanger. Desalinating ducks will have shorter water-line
lengths than electricity-generating ones. They will move through large angles
in quite small waves. There is an interesting optimisation equation which
takes input from wave climate statistics, duck hydrodynamics, steam tables
and the costs of heat-transfer surface. It looks as though ducks should
be designed for power levels in the range 3 to 10 kW/m with hull diameters
of 6 or 7 metres. Solo ducks can extract power equal to twice that in their
own width of sea and work well in wavelengths forty times their diameter.

The vapour compression process produces one part of fresh water and one part
of double-strength brine for every two parts of sea water fed in. But both
output streams are at a temperature of about 100°C and this heat must be
transferred to the input stream. The heat-exchanger to do this will draw
a very particular advantage from being mounted in a floating device. The
arrangement is sketched in Fig. 3. Four long sheets of heat-transfer surface
are wound in parallel on to a central hub, which contains connecting
manifolds. The sheets are separated at their edges by a pair of elastomeric
strips which are compressed by sheet tension and by the clamping of an outer
ring. There are no welds and no metallic contacts.

The pressure required to force fluid through a very long spiral path would
be large for the sealing method described above. But if the assembly is
mounted with its axis on the nod axis of a duck the entire pumping action
can be induced directly by the duck motion. The spiral of fluid will stay
nearly still while duck nodding moves the heat-transfer surfaces vigorously
past it. No pressure need ever appear. All that is necessary is to have
a texture formed in the sheets which has a fractional difference in flow
resistance in the two directions. This will apply a continuous distributed
pumping effort along the full length of the heat exchanger which will balance
the resistance to flow. It is difficult to think of a more satisfactory
heat-transfer mechanism. The heat-exchanger cost should not be a large
multiple of the cost of the raw material.

THE SCALE PROBLEM

Some of the innumerable chemicals in sea water are close to their solubility
limit and some have solubilities which fall with rising temperature. This
makes them precipitate on to the nearest substrate and the effects on heat
transfer are disastrous. Scale weight can reach 1/1000 of product weight
so that a medium size installation may be getting a tonne of scale a day.

G1.4

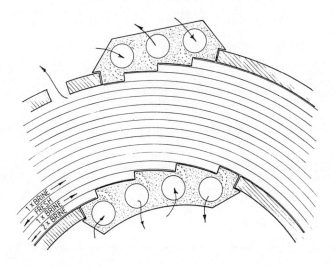

Fig. 3.   An axial view of the nod-driven secondary heat
          exchanger, which pre-heats the feed water

The usual remedies are treatment with acid or more exotic chemicals, combined
with scraping and thermal shocks. The acid methods can induce severe internal
corrosion so that expensive materials may be necessary. The prospect of
doing difficult chemistry on a structure designed for violent response to
waves seems most unattractive. The scale problem is so serious that it seems
we should entirely forget both thermodynamics and hydrodynamics and
concentrate on the development of a long-life, scale-free process. I believe
that this may be best achieved by using a heat-transfer medium developed
by my colleague Maxwell Davidson. It will shortly be available from James
Howden Ltd. of Glasgow. The current cost is £15 per square metre.

The chemical scales which form on heat-transfer surfaces are hard and brittle.
They can stick to rigid metallic plates with considerable tenacity. But
both the rate of deposition and the strength of adhesion on plastic surfaces
is very much lower. Davidson's idea is the use of mesh woven from a
reasonably good conductor covered with a very thin, say 0.2mm, layer of heat-
resisting plastic. The mesh provides a strong substrate for the plastic
and its interstices allow a more positive bond than would a smooth surface.
The mesh also conducts heat easily in the plane of the composite so that
local hot-spots are avoided. The plastic isolates the metal from corrosive
influences and offers the scale an unattractive substrate in the manner of
the non-stick frying pan. Finally the combination can flex more easily than
can the layer of scale so that there is an additional scale removal mechanism
which so far seems to have been little exploited. Unfouled mesh-plastic

surfaces transmit about 80% of the heat of comparable stainless steel
specimens. In a 72 hour test in contact with boiling brine of double sea-
strength the weight of scale deposited on the plastic sheet was only 1/60th
of that on the stainless steel. Furthermore its adhesion was so slight that
it could be shaken off. If these features can combine with controlled flexure
it may prove possible to make long-life, scale-free heat-exchangers.

If the assembly is fully flooded and in a rigid casing, the lateral
accelerations from wave motion will pass through the fluid and there need
be no stresses in the heat-transfer surfaces. This would also be true if
the casing contained fluid-filled bags. However, if fluid in these bags
were to be replaced by air, then lateral accelerations would deflect the
sheets of heat-transfer material to a limit allowed by the air volume. This
means that a deflection pattern can be deliberately introduced for the
purposes of scale removal. Particles thus removed will form an alternative
nucleation site for other deposits and an abrasive to scour the heat-transfer
surfaces. Some careful thought must be given to the frequency of the
descaling operation so as to avoid excessive scouring action.

SECONDARY PUMPING

Vapour compression systems need all sorts of pumps to move water through
various networks and remove dissolved gases. We want to eliminate the
electric motors and their control equipment which are usually used. A helical
loop of pipe with non-return valves at each end can induce positive and
negative pressures equal to the product of the density of the fluid, the
total length of the coil, the radius of the loop and the angular acceleration
of the duck. A small number of loops can raise all the pressure needed for
sending the product to an elevated inshore reservoir and all the vacuum needed
for gas removal.

Solo ducks obtain their reference through tension-leg moorings to the sea
bed. The mooring lines can be used to perform two other useful functions.
The first is to assist the secondary heat-exchanger. It is not possible
to discharge the output streams at exactly ambient temperature and so valuable
heat will be lost. It looks as though a small amount of heat-pumping can
play a useful role. The variations of tension in the mooring lines can be
used to drive Freon pumps for this purpose and so a secondary heat-exchanger
of modest area can enjoy a larger temperature gradient. The amount of heat
lost through the insulated hull (which turns out to be surprisingly small)
can be replaced and the process started easily from cold.

A further function of the mooring lines concerns the bearings of the duck.
These are spherical journals with a diameter of one metre, fitted inside
a rubber outer. Fully hydrostatic operation is ensured if they are fed with
part of the fresh-water product stream pressurised by pumps in mooring lines.

Both the standing tensions and, more importantly, the variation in tension
in the mooring lines of a solo duck are very much higher than those of the
close-packed spine-mounted version. We expect a variation of ±2.5 meganewtons
on a mean level of 5 meganewtons. Very careful attention must be given to
the terminations and the attachment to the sea bed. Some relief can be
provided by flooding the hull (preferably with fresh product water) and
submerging the duck in extreme conditions. Paradoxically a small depth of
submersion would give it a better chance of surviving a hurricane than land-
based plant.

PRODUCTIVITY

Fresh water output depends more on swept angle than on absorbed energy.
At the cost of increasing the amount of heat exchanger area we can improve
the energy efficiency of the process. At present we believe that the optimum
area would require between 20 and 40 kilojoules per kilogram of product (about
5 - 10 kilowatt hours per tonne). This approaches the conversion efficiency
of the best reverse osmosis plant and would mean that a 25 metre wide duck
would produce as much as 10 kilograms per second (over 800 tonnes per day)
in a sea of 10 kilowatts per metre. Output would fall with the square root
of power absorbed.

If installation and the laying of the pipe to shore can be efficiently
organised, it should be possible to get the capital cost below £500,000 per
duck. The resulting water costs may be attractive in many parts of the world.

OTHER METHODS

In parallel with vapour compression we have considered wave-powered reverse
osmosis. A solo duck can obtain a torque reaction from its tension leg
moorings and generate power in the form of high-pressure oil by means of
ring-cam pumps. This energy can be converted to the pressures suitable for
reverse osmosis by reciprocating-piston transformers, so that the very minimum
of the pumping plant is exposed to salt water. The initial cost estimates
are slightly in favour of the vapour compression method because of the very
cheap pump. A reduction in the cost or improvement in the life of reverse
osmosis membranes could close the gap. It would be wrong to omit a reference
to the pioneering work of Hicks and Pleass (1985) who use heaving-floats
to drive reverse osmosis capsules. The stringent pre-filtering requirement
is performed by a sand-well using the sea bed itself. Six units have been
in operation off Puerto Rico since 1982.

ACKNOWLEDGEMENTS

Wave energy research at Edinburgh University is supported by the United
Kingdom Department of Energy. I am particularly grateful to Maxwell Davidson
for teaching me about vapour compression and for the suggestion that it could
be applied to Wave Energy.

REFERENCES

Davies, P.G. (1985). Wave Energy - The Department of Energy's R & D Programme
    1974-83. ETSU R26. HMSO, London.

Hicks, D.C., and C. M. Pleass (1985). Physical and mathematical modelling
    of a point absorber wave-energy conversion system with non-linear damping.
    In A. F. de O. Falcao (Ed.), Proceedings of IUTAM Conference on
    Hydrodynamics of Ocean Wave-Energy Utilization, Springer-Verlag, Berlin.
    (In print).

Porteous, A. (1983). Desalination Technology - Developments and Practice.
    Applied Science Publishers, London.

Salter, S. H. (1985). Progress on Edinburgh Ducks. In A. F. de O. Falcao
    (Ed.), Proceedings of IUTAM Conference on Hydrodynamics of Ocean Wave-Energy
    Utilization, Springer-Verlag, Berlin. (In print).

The Wave Energy Module

H. H. Hopfe* and A. D. Grant**

*U.S. Wave Energy, Inc., Longmeadow, MA, USA
**Department of Thermodynamics and Fluid Mechanics,
University of Strathclyde, Glasgow, UK

ABSTRACT

A wave energy conversion device, capable of producing electricity for as little as 2 cents per kWh, is described. The system uses proven components and has been the subject of exhaustive model tests at 1/10 scale. Plans for marketing and deployment of the first 1 MW production prototypes are at an advanced stage.

KEYWORDS

Wave Energy; modular system; electrical power generation; flexible marketing strategy.

DESCRIPTION OF THE DEVICE

The Wave Energy Module or WEM consists essentially of two parallel discs connected with hydraulic piston pumps. One disc is buoyant and rides on the ocean surface, while the other (the reaction plate) is suspended at a depth where motion due to surface waves is small. The layout of the system is shown in Fig. 1. Hydraulic fluid is transferred from the piston pumps to a high-pressure accumulator. It is then fed to a hydraulic motor connected to a generator. Accumulator, motor and generator are all housed on the upper disc (the raft). A typical layout of the hydraulic circuitry is shown in Fig. 2.

DEVELOPMENT HISTORY

The working principles of the device were established in model tests at 1/40 scale, conducted at the Alden Research Laboratory of the Worcester Polytechnical Institute and at the University of Massachusetts, which began in 1976. In 1979, a series of performance and storm survival studies were initiated on Lake Champlain near South Hero, Vermont with a 1/10 scale prototype fitted with a 1 kW electrical generator. These tests successfully demonstrated the production of electricity by a WEM and the storm survivability of the module. Energy conversion efficiencies as high as 93% were observed. These tests were successfully concluded in September 1983.

G2.1

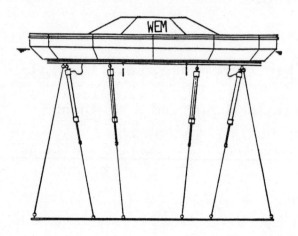

Fig. 1.   Conceptual model of 37m diameter, 1 MW WEM

SYSTEM SCHEMATIC

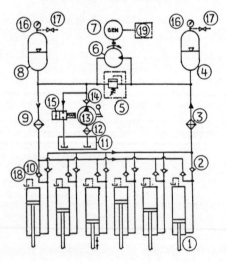

COMPONENTS

1. PRIMARY PUMPS
2. SUPPLY CHECK VALVES
3. INTEGRAL OIL COOLER
4. HIGH PRESSURE ACCUMULATOR
5. PILOT OPERATED RELIEF VALVE
6. HYDRAULIC MOTOR
7. GENERATOR
8. LOW PRESSURE ACCUMULATOR
9. FILTER
10. RETURN CHECK VALVES
11. HYDRAULIC RESERVOIR
12. STRAINER
13. MANUAL CHARGING PUMP
14. CHECK VALVE
15. MANUAL DISCHARGE VALVE
16. TRANSMITTING PRESSURE GAGE
17. AIR CHARGING VALVE
18. PRIMARY PUMP VENT
19. GENERATOR FIELD CONTROL

Fig. 2.   WEM hydraulic/electric system schematic

G2.2

Details of the WEM have been presented at the 1st BHRA Symposium on Wave and Tidal Energy (Hopfe, 1978) and at the 1981 IEEE Oceans '81 Conference. It has been the subject of articles in Science Digest and Design News Magazines, and has received coverage in Current Affairs programmes in the US National Television Network. From a large number of potential candidates, the WEM design was chosen in 1984 for display at the US Patent Office in the Thomas Edison Exposition.

A company, US Wave Energy Inc., has been formed to promote commercial development of the WEM. Two designs for larger prototypes have been prepared: by General Dynamics in 1981 for a 30 ft (9 m) machine, and by Giannotti and Associates in 1984 for a 1 MW prototype, to the same scale as the production version.

TECHNICAL ASPECTS OF WEM PERFORMANCE

The WEM is essentially a two-component system, power being produced by the relative motion of these two components. In this respect and in the power take-off system employed, it resembles the wave-contouring raft developed by Wavepower Ltd. in the 1970's (Cockerell et al, 1978). The geometry of the WEM is of course completely different, and it is this which gives it a number of significant advantages:-

a)   it is omnidirectional, operating with equal efficiency for all incoming wave directions

b)   power is produced by relative motion in any of the six possible degrees of freedom between the raft and the reaction plate

c)   the power produced in storm conditions is self-limiting. As the wave size increases, the water in the vicinity of the reaction plate becomes influenced by motion on the surface. The reaction plate begins to follow the movement of the raft, so limiting the power developed and, more importantly, limiting the structural loads on the system. This aspect of WEM behaviour has been verified by the model tests on Lake Champlain. The vertical space between raft and reaction plate is of course critical in determining the maximum power which a given WEM can produce, so a system may be designed to give any predetermined maximum power value.

In addition to these features, the fact that the WEM is a de-tuned system, i.e. it does not rely for operation upon any resonant behaviour, means that it can capture energy over a wide range of wave periods with high efficiency. This efficiency will be maximised in production systems by arranging for the hydraulic fluid pressure to be adjustable. A wave monitor some distance from the raft will send data to an on-shore facility, where the optimum fluid pressure value is computed and the necessary control signal sent to the WEM.

The raft and reaction plate of the WEM are required to be stiff but not massive and material costs should therefore be moderate. This is a "low technology" system using long-established methods and components with proven service records. Mooring is not now considered to be a major

problem; large discus buoys are commonplace, and while the WEM at 37m
diameter is a lot larger it is small in relation to many other wave energy
conversion devices proposed. Flexing of the power cable is inevitable as
it is attached to the raft and the latter is of course intended to move in
response to the waves. Studies in this area are continuing.

Regular maintenance of the WEM is envisaged at two-year intervals. After
this period, replacement of piston pump shaft seals would be a sensible
precaution, and this would be done in practice by withdrawing the complete
piston unit and replacing it with a new one. Inspection and maintenance
of the equipment on the raft would be done at the same time.

APPLICATIONS

If it is accepted that an annual mean of 20 kW/m is an acceptable minimum
wave regime for energy conversion, it is seen (Fig.3) that wave energy is
a very widespread resource. A comprehensive review of the possibilities
for exploitation is given by Quayle and Changery (1981). Most of the
proposals made hitherto have been for the deployment of large
multi-megawatt units connected to national power supply networks.

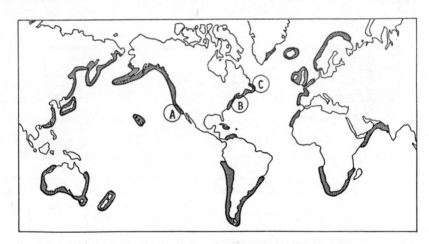

Fig. 3. Global distribution of wave energy exceeding 20 kW/m

Nowadays however, ideas are increasingly turning towards the use of
smaller systems to fulfil local needs. Fig. 4 summarises the areas in
which the WEM might make a contribution. The most attractive of these are
where deployment would involve the supplementing or replacement of diesel
generating plant, in coastal or island communities where there is no
connection to a national electricity grid. The relatively modest size and
output of the WEM (in comparison to many other proposed devices) makes it
particularly attractive in this application. Increasing activity around
the coastlines of the world, in exploration, mining, oil and food
production, points to a growing demand for power. The U.S. Department of
Energy estimates that the demand around its coasts alone could exceed 12
GW by the 1990's. Here again, the WEM is well placed to make a
contribution.

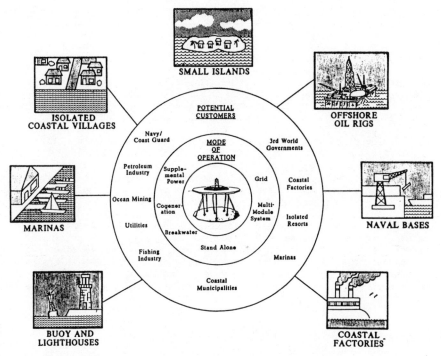

Fig. 4.   Projected applications for WEM

PROJECTED COSTS

The proposed production unit (37 m diameter, 1 MW rating) is costed at
$1.8 M to construct and install. The mean annual output will of course be
much less than 1 MW and there will be maintenance costs to be met.
Detailed estimates of energy costs are possible for specific sites where
the wave climate is well known.   Based on a system life of 20 years,
figures in the range 2 to 8c/kWh are produced. These compare very
favourably with costs for diesel generation (say 20c/kWh) and would
suggest short pay-back times.

U.S. Wave Energy Inc. have developed a marketing strategy, outlined in
their Business Plan (Hay and Werner, 1985) which presents the customer
with a number of options. As an alternative to outright purchase of the
WEM, the customer may contract to purchase electricity at a negotiated
price from a consortium composed of U.S. Energy Inc. and investors. An
option for subsequent purchase of the WEM may be retained by the customer
if desired.   In this way, risk to the customer, an obvious cause for
concern until the device proves itself in operation, is minimised.

PROPOSALS FOR FIRST DEPLOYMENT

Three serious proposals for deployment of the first WEM are presently
being negotiated by U.S. Wave Energy Inc. The locations are shown in
Fig.3.

At San Clemente Island, California [location (A)], the U.S. Navy is
seeking bids from interested parties for an electrical power production
system based on a renewable resource. The location is such that wave
energy is an obvious candidate. Their present system uses diesel
generators, to supply a peak load of about 850 kW. U.S. Wave Energy Inc.
have put in a formal bid for this project.

At Nantucket Island [location (B)], the utility company Boston Edison is
seeking to supplement or replace the diesel-driven generators presently
used. Wind and wave energy are both being considered. About 15 WEM units
would be required to meet peak power demand on the island.

The Newfoundland Oceans Research and Development Corporation (NORDCO) has
identified a number of communities on its South coast [location (C)] which
presently rely on diesel-generated power for consumption domestically and
in fish processing plant. NORDCO has expressed a strong interest in the
use of the WEM in this region, and would be prepared to manage the entire
operation. Discussions on funding are now in progress.

It is only natural that these initial proposals should refer to territory
either within or close to U.S. coastal waters, given the origins of the
parent company. However, interest is growing in a number of other regions
throughout the world. The West coast of Scotland, with its island
communities, its high-energy wave climate, and its proximity to
under-employed oil platform construction facilities, is a region where the
WEM concept should be particularly attractive.

REFERENCES

Cockerell, Sir C., M.J. Platts and R. Comyns-Carr (1978). "The
development of the wave contouring raft. Proceedings of the ETSU, Harwell
Laboratories Conference on Wave Energy, Heathrow Hotel, London, England,
pp 7-16.
Hay, G.A., III and E. Werner (1985). "U.S. Wave Energy Business Plan".
Hopfe, H.H. (1978). Extraction of ocean wave energy by means of a
constrained floating platform - power efficiency study I". Proceedings of
BHRA International Symposium on Wave and Tidal Energy, P S1-1, University
of Kent, Canterbury, England.
Quayle, R.G. and M.J. Changery (1981). "Estimates of coastal deep water
wave energy potential for the world". IEE Publication Number 81CH1685-7,
Oceans 81 Conference Record (vol.Two), p8 903-907.

Wave Energy Power Stations in Norway

K. Bønke

Norwegian Hydrodynamic Laboratories, Trondheim,
Norway

ABSTRACT

An account of the world's first commercial wave power devices for grid connection.   Mostly describes the 500 kW multiresonant oscillating water column system.   Includes estimates of the electricity production costs.

KEYWORDS

Wave power, Norway

INTRODUCTION

Two different types of wave energy stations are now being built in Norway. These projects are occuring at the end of a $15 M research and development programme and at the start of a commercial period. The programme so far has been financed in cooperation between the Norwegian Government (OED) and private industry.   In this paper the so-called Kvaerner's multi-resonant oscillating water column system (MOWC) is described. (See also paper G1).

Although wave energy is globally a small source of energy, it has the possibility of a significant impact on many coastal and island communities. Historically the availability of a clean, low-cost and renewable source of energy, namely hydro power, has been one key factor in Norway's economic growth.   Wave energy may have a similar effect on communities with long coast-lines per capita.

The two prototypes are quite different in design, but have the common feature of being mechanically simple and rugged.   One fact which came out of the R & D programme on wave energy both in Norway and UK, was the need to make wave power stations rugged and simple. Maintenance cost would otherwise become prohibitive.

KVAERNER MULTIRESONANT DEVICE

(1)   The multiresonant oscillating water column (MOWC)

One of the prototypes being built is a so-called oscillating water column.

G3.1

249

Here, a "water piston" driven by the incoming wave, pumps air back and forth
- through a symmetrical turbine. The Norwegian design is characterised by
two important benefits:

(a)   The harbour

The device has has two protruding walls in front of the column's
opening. These walls form the "harbour".   This harbour forms a
quarter-wave resonator. It is therefore possible to select dimensic
so the deivice has two resonance frequencies - one for the column
and one for the harbour.

Wave energy absorbers of this kind absorb energy only from waves
with a period close to the absorber's resonance period. Introducing
more than one resonance frequency makes it possible to approximately
double the energy production.   Therefore the resonant harbour is a
very cost-effective device.   Initially the idea was to have the
absorber continuously tunable for resonance. However several facts
obtained in model testing made us abandon the idea in favour of a
fixed harbour configuration.

(b)   Point absorber effect

Wave energy absorbers working at resonance will absorb energy from a
crest-length longer than the absorber length (along the crest).

The combination of these two facts comes out with the net result that this OWC
yields more kilowatthours per ton material used to build it than any other OWC
concept   presently under development.

(2)   Turbine design

The air turbine used in the OWC prototype rotates in the same direction
irrespective of the air flow direction, Fig.1.   The original concept was
developed at Queen's University in Belfast by Prof.Wells.   The concept has
been further developed by Kvaerner, the company behind the OWC prototype.
(Kvaerner is a large manufacturer of turbines, and offshore constructions
among other things).

Originally the idea was to place the oscillation chamber structure in the ope
sea, anchored to or resting on the sea bottom.   Civil work in an offshore
environment is however an expensive undertaking.   The site finally chosen
for the prototype is an almost vertical wall of rock on a peninsula.   This
made the building of the power plant a cost saving onshore operation.   The
main task was rock blasting, which seems to be the cheapest part of civil
works.   Also the access during mechanical installation, testing, inspection,
maintenance and demonstration is easy and not influenced by weather conditior

The wall of rock facing west falls nearly directly to 60 m below sea level.
Hence westerly waves travel almost undisturbed up to the plant.   In addition
the wall acts as a reflector, a fact which increases the power absorbtion
significantly (Fig.2 and 3).

The "opening angle", ie. the angle within which waves travel undisturbed, is
approx. 70° from north west to south west.   This is a rather narrow opening
angle, but the site is selected as a compromise between easy access and
favourable wave conditions.   The western coast of Norway has, however, a

multitude of places with more favourable wave conditions.

(3)   The layout of the plant (Fig.4)

The oscillation chamber has a cross section at sea level of about 50 m^2.
Vertical dimensions allow oscillations of maximum ± 3.5 m.

The air turbine is placed 15 metres above sea level on a steel plate tower.
The inlet/outlet opening of the turbine is controlled by a hydraulically
operated cylindrical gate.   A specially designed barrier protects the
rotating machinery against splashing water without disturbing the air flow.
The turbine is designed to have a maximum instant capacity of 1,000 kW or a
maximum average capacity over a wave period of 500 kW.   The air turbine
always rotates in the same direction regardless   of the direction of the
airflow.   Through a flexible coupling the turbine is connected to an a.c.
generator with a capacity of 600 kVA at 1,500 rpm.   The generator is
connected to the grid over an ac-dc-ac converter.

PRELIMINARY RESULTS

The civil works were completed in the early autumn 1984.   The steel tower
was erected later the same year.   Instead of the turbine, a controllable
valve simulating the turbine was placed on top of the tower.   The valve was
replaced by the turbine and the generator in September 1985.   Utilizing
this valve arrangement, a measurement and test programme was conducted.

Information of incoming waves and corresponding pneumatic power production
were collected and evaluated (Fig.5).   Measurements of pneumatic energy
production as a function of the energy flux per meter wave front are
convincing.   They correlate well with predictions based on theoretical works
and model testing done during the R & D phase at the Norwegian Hydrodynamic
Laboratories (Fig.6).

ENERGY COST

The cost of producing one unit of energy depends mainly on two factors.
First, energy production which depends both on the efficiency of the wave
power plant and the wave energy flux at the site.   Second, the cost of
building and installation which depends on the site and even more so on
the simplicity of the power plant.   The cost estimated has changed dramatically during the years from 1980 to 1985 (Fig.7).

Based on the experiences from the Kvaerner wave power plant so far, cost
may be predicted.

(a)   One power plant subsequent to the one which is now being built,
using the same design and the same sub-suppliers, would cost an
additional $600,000 (US).   Assuming a yearly average energy flux
of 15 kW/m of wave front would give a cost of $0.40 kW/year.
Dependent on interest, depreciation rate and real maintenance cost,
the cost of the wave energy will be $0.04-0.05 per kWh.

(b)   Building 10 units in the same wave climate and with similar
site conditions as for the prototype case, would lower the energy
cost to $0.03-0.04 per kWh.

CONCLUSION

The building and testing of the prototype may reveal weaknesses.   However
new ideas are already developing concerning both efficiency and simplicity in
design.   Combined with the positive results from the prototype testing so far
we feel confident in the future of this wave power device.

The main message of this report is that wave power stations in the 500 kW range
are being built.   These will be connected to the grid for generation in
autumn 1985.   Power prices are expected to be competitive, particularly in
regions which today depend on diesel generation but have wave energy available

REFERENCES

Ambli, N, Bonke, K, Malmo, O. and Reitan, A. (1982).   "The Kvaerner multi-
    resonant OWC"   Proc.2nd Int. Symp. on Wave Energy Utilization, Trondeim
    Norway, June 22-24, pp 275-295.

Evans, D.V. (1982).   "Wave-power absorption by systems of oscillating surface
    pressure distributions".   Journ. Fluid Mech. $\underline{114}$, pp 481-500.

Evans, D.V. (1982).   "Wave-power absorption within a resonant harbour".
    Proc.2nd Int. Symp. on Wave Energy Utilization.   Trondheim Norway, June
    22-24, pp 371-378.

Malmo, O. (1984).   "A study of a multiresonant oscillating water colum., for
    wave-power absorption", PhD thesis.   The Norwegian Institute of Technology
    Department of Physics, University of Trondheim.

For further information please contact:

        Kvaerner Brug A/S
        Wave Power Project
        P O Box 3610 Gamlebyen
        N-0135 OSLO 1

        Phone: +472 68 55 50
        Telex 71650 kb n

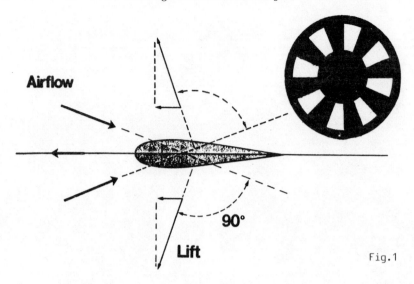

Airflow

90°

Lift

Fig.1

KB's M.O.W.C. at
Toftestallen
nr Bergen

Fig.2

G3.5

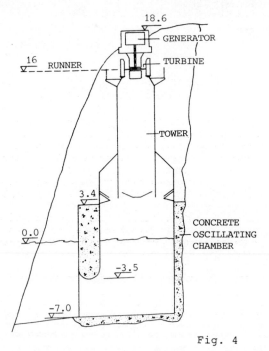

Fig. 4

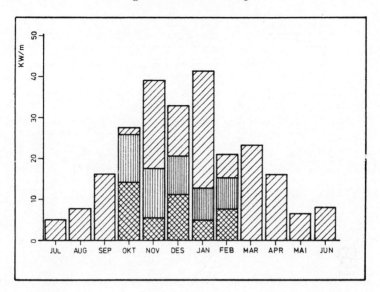

Mean energy flux, Utsira 1976 – 1984
Mean energy flux, Utsira 1984 – 1985
Mean energy flux, Toftestallen 1984 – 1985

Fig.5

Kværner Brug A/S

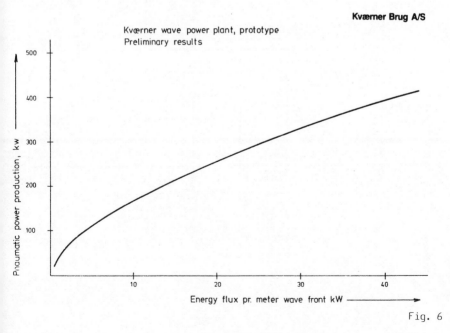

Fig. 6

G3.7

256 K. Bønke

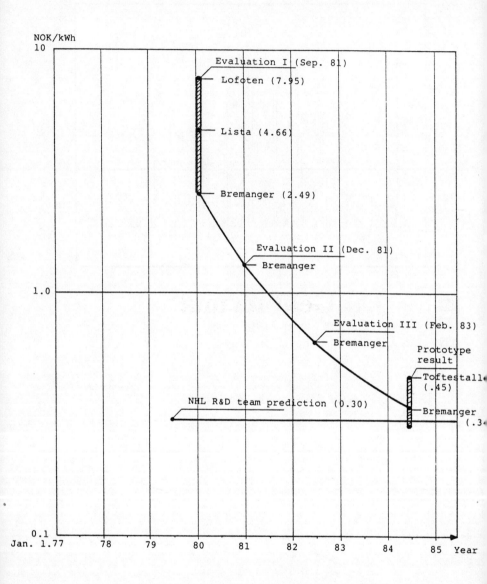

Fig.7

G3.8

TOPIC H
Geothermal

Energy Source for Agroindustrial Integrated Systems

D. R. Mendoza, Jr., R. O. Bernardo and H. V. Guillen

Engineering & Construction Department, Geothermal Division, Energy Development Corporation, Philippine National Oil Company, Merritt Rd., Ft. Bonifacio, Metro Manila, Philippines

ABSTRACT

The Philippine Islands possess vast reserves of geothermal energy, some of which are situated in very remote areas or in islands wherein only traditional agricultural and handicraft industries exist.

This paper examines how geothermal heat can be used in agroindustrial systems for post harvest processing of food and commodity products. These systems can utilize a cascaded type of small scale industrial streams and power generation on a limited scale.

Economic and social benefits of this developmental approach is highly changeable. Developers should be cautioned to exercise synergism on a micro-regional approach on infrastructure, agriculture and rural development programmes.

The model used in the study is Manito, an agricultural town in the province of Albay which has a population of just over 15,000. The area is considered as one of the economically depressed areas, where 70% of the population earn US$210 per annum as of 1985. The implementation of an agroindustrial system for communities such as this model are expected to have an immediate positive socio-economic outcome.

KEYWORDS

Synergism; single flash power generation; agroindustrial integrated system; binary cycle power plant; triple-effect evaporator.

PROJECT BACKGROUND

Geothermal potential in the Philippines remains unmeasured. Power generation now exceeds 880 MW while direct process use of geothermal fluid is virtually nil and salt recovery from sea water remained experimental since 1967. Some of the existing geothermal fields that have been developed for power gene-

ration can be found in the forested regions of the island group of the
Central Philippine Visayan islands. A number of exploration wells were
drilled in the lowlands, with low to medium temperature, while higher
temperatures (in excess of 170°C) were being sought for use in single flash
power generation schemes. It is now believed to be appropriate to consider
the lowland fields for non-electrical use of the fluid.

Exploration works oftentimes lead to the discovery of geothermal fields from
low to higher temperatures. The lower temperature reservoir may be found on
lower grounds towards the coastal areas, surrounded by vast agricultural
plains and accompanied by thriving fishing industry. Traditional handicraft
industries based on hemp fiber materials are usually on full operation in
between harvests. As it is historically the case of the farmer praying for
rain to cultivate his land, he loses the benefit of the sun to dry his
produce. The area of interest is particularly susceptible to extremely
severe weather conditions caused by seasonal typhoons. Almost 50% of typhoons
crossing the country affect the area. Based on recorded statistics, there
is no pronounced dry season and large amounts of rain usually fall throughout
the year. It cannot be more timely for geothermal heat to play an important
role in the lifestyle of the farmer to provide drying processes, and extend
to a full-scale agroindustrial complex. The model for this study is Manito
Albay in the southern part of the main island of Luzon.

This southern region, although bountiful of marine resources and coconut
groves, has an annual per capita income of 20% below national average and
unemployment rate twice the national level. And yet, it is still not the
most economically depressed area in this Islands Nation. Therefore,
development projects bearing great influence to the agricultural sector are
very attractive.

Manito land area has an immediate coverage of 26.5 square kilometers and
another 80.9 sq. kms. within 40 kms. of road distance with rice as the basic
crop. Marine biology of the adjacent Poliqui Bay is quite rich and fairly
diverse community. The Simpson's diversity index is 0.86 which alone
represents 11% of known Philippine fish species. The Albay Gulf has an even
higher diversity index of 0.94 while 95% of the marine species are edible.
This is already substantial to support a fishing industry for a rural
population of over 15,000.

The figures below show the population distribution by major source of
livelihood.

TABLE 1   Livelihood Classification as of 1982*

Source	Distribution
Farming	72.1%
Fishing	3.6%
Manufacturing	8.8%
Retail Trade	2.0%
Salaried Occupation	10.8%
Miscellaneous Sources	2.7%

*Figures were derived from the records of the
National Census and Statistics Board.

The distribution may just be right except for the extremely low figure for fishing. This is however quite explainable that inspite of extremely potent marine resource, few indulge other than for local consumption because no facilities are available to support this industry based on highly perishable produce. Farming, having the highest contribution, relies heavily on sun drying of the grain just as those involved in manufacturing of rope fibers. This processing based on ancestral procedures influence to a very great extent the actual output of the land and sea which ultimately keep the income at a very low level.

Post harvest processing can therefore be developed based on existing geothermal heat. This Agroindustrial Integrated System can sustain the Manito area and its environs. A similar module may be built in a nearby town or even in another island.

DEVELOPMENT PROGRAMME

A total energy integrated system can be engineered utilizing geothermal steam cascading from the first industry requiring the highest temperature like a refrigeration system to support the fishing industry down to an industry requiring very little heat, like aquaculture.

Before an Agroindustrial Integrated System can be engineered, an agribusiness flow pattern requires a detailed investigation to identify the size of processing industries that are to be built. Infrastructure programmes need to be rationalized with development schedules and to be studied carefully for the impact on the lifestyle of the rural folks.

For the town of Manito, the following annual figures were collected as baseline data for the produce. Extreme difficulties may be encountered if similar studies are done in developing and least-developed countries.

TABLE 2   Volume of Domestic Products Processed in 1982

Produce	Area/Population	Quantity Processed
a. Temporary Crops		
Rice	631 Hectares	1817 M. Tonnes
Corn	80 Hectares	88 M. Tonnes
Vegetable	132 Hectares	498 M. Tonnes
Rootcrops	238 Hectares	774 M. Tonnes
b. Permanent Crops		
Coconut	1,168 Has./168,800 Trees	3,553,000 Nuts
Copra		79 M. Tonnes
Coffee, cacao, fruits	71 Hectares	287 M. Tonnes
Abaca		No available data
c. Poultry		
Chickens, ducks, turkeys, geese	19,640 Birds	0.41 M. Tonnes
d. Livestock		
Cattles, carabaos, hogs, goats	7,140 Heads	67 M. Tonnes

Produce	Area/Population	Quantity Processed
e. Fishing		
Tuna, mackerel, shells, etc.	453 Boats	95 M. Tonnes
f. Forestry		
Timber	3,880 Hectares	106,000 cu.m.
Firewood		11,140 cu.m.
Bamboo		502 pieces
Rattan		3,670 meters
Palms		5,500 pieces

The development programme envisioned for the Manito Agroindustrial Estate is
staged for a period of 15 years, as shown figuratively below.

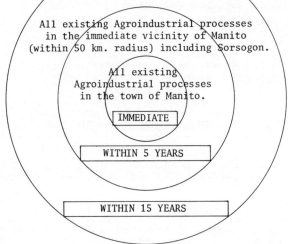

Introduction of non-traditional industries in Manito and
its suburban localities relocated from other areas or
industries based on country development programme.

All existing Agroindustrial processes
in the immediate vicinity of Manito
(within 50 km. radius) including Sorsogon.

All existing
Agroindustrial processes
in the town of Manito.

IMMEDIATE

WITHIN 5 YEARS

WITHIN 15 YEARS

Fig. 1.   Stages of geothermal energy utilization for
agroindustrial applications in Albay and
Sorsogon provinces

A policy is now needed for the government to adopt for purposes of coordinating
the pricing structure of geothermal heat, value added and marketing of produce
government taxes, subsidies and incentives.

This last part is perhaps the more complex stage of the study as this involves
different sectors but at the same time preserve the local farmer by establishing a cooperative to be managed by the local farmers themselves.

This development programme is presently being investigated in Manito with the
assistance of the United Nations Development Programme for possible development
in the near future to benefit the rural population and its environs.

GEOTHERMAL PROCESS

Geothermal fluid, comprising of 30% steam and 70% water, discharging from the production well is transmitted to the separator vessel through the two-phase pipeline. The fluid is then separated into steam and water by centrifugal force inside the vessel.

The water leaves the bottom of the vessel at 150°C and is fed to the Binary Power Plant, which in turn generates the electric power requirement of the agroindustrial complex. Likewise, geothermal water at 150°C is fed to process fish for drying. The excess hot water from the Binary cycle is further utilized by the fish drying plant for dehydration process. Finally, the disposal of waste water is done by discharging it to a settling pond to separate calcium silicate by gravity before transferring to a holding pond for further cooling.

Similarly, the separated dry steam leaves through a central pipe emerging from the bottom of the separator and is then distributed to various drying plants. The steam has a quality of 99.95% and total dissolved solids of less than 10 ppm.

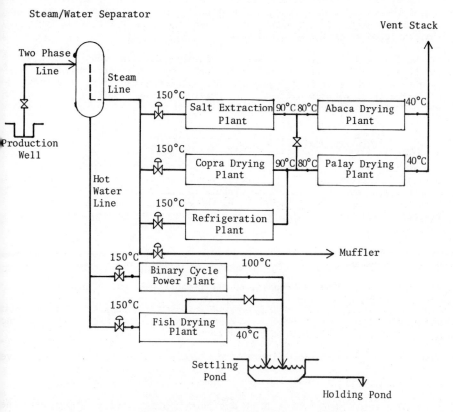

Fig. 2. Geothermal process flowchart for the proposed
         Agroindustrial Complex

The primary users of the separated steam are the Salt Extraction Plant, the Copra Drying Plant and the Refrigeration Plant. They have a relatively high temperature requirement to generate and produce the desired finished products. The downstream plants, however, comprise of Abaca fiber and Palay grain drying which require a much lower temperature of steam. The exhaust steam is then discharged to the atmosphere through a vertical stack.

The system is fully protected from any abnormalities brought up by the upstream and downstream plants. It is equipped with numerous safety devices and the overpressure control system always has a spare unit.

SUMMARY

The direct utilization of geothermal fluid for agroindustry varies from one country to another due primarily to climatic conditions. The Philippines, being a tropical country can only confine itself in the processing of resources abundant in the islands. Unlike in temperate countries, utilization can extend up to space heating and industrial chemical processing to name a few.

The model in this paper illustrate the kind of industries to which the direct use of geothermal fluid may apply locally. The heat exchanger, which oftentimes replaces the conventional fossil fuel drying equipment, constitutes the main element of the plants design.

If the design of the heat exchanger is someday perfected, most particularly in the aspect of material selection and efficiency, a great economic advantage can be envisioned enhancing the upliftment of the socio-economic condition of rural communities in the Philippines and elsewhere in the world where geothermal resource is available.

ACKNOWLEDGEMENT

The authors would like to express their gratitude to the NCSO Region V Director for providing all the necessary records and statistics, the National Economic and Development Board for furnishing the much needed directories of complementing agencies in Region V and the Bicol River Basin Development Project Office for providing programme of development in Albay and Sorsogon provinces.

REFERENCE

Alcaraz, A. P. (1979). A Proposal on the Establishment of an Industrial Estate Centered Around the Manito Geothermal Area.

Italian Geothermal Projects for Rural and Island Communities

R. C. Lesmo

ONG (Ordine Nazionale Geologi), via della Conciliazione
22, Rome, and via J. F. Kennedy 16, 20097 San Donato
Milanese, Milan, Italy

ABSTRACT

The main Italian geothermal projects are reviewed according to the geographical
and geological situations .

Prevailing rural projects were developed or under development in the "hot"Tyrrhenian belt ; projects for district heating but also for joint civil and rural
aplications were developed or are under development in the "cold"Adriatic-ionic
belt around spas or where dry petroleum wells found promising intermediate
temperature geothermal fluids . Outside the peninsula the Vulcano project for
joint fresh water and electricity production and a few projects in Western
Sardinia stress the interest of geothermal energy for island communities.

The new agriculture philosophy asking for renewable energy sources and appropriate technologies is in favour of direct geothermal applications but geo-heat
utilization may result conditioned by geographical, geological, social and political conditions . Appropriate evaluation and management of geothermal projects may develop a wider utilization of local resources for thermal rural
applications where electrification is most expensive and elsewhere through
heat and utilization integration .

KEYWORDS

Geothermal province ; geothermal project; log-normal distribution ;prospect;
preevaluation ; evaluation ; geothermally oriented territory planning .

ITALIAN GEOTHERMAL PROVINCES AND PROJECTS

High temperature geothermal fluids are economically recoverable mainly in regions of recent volcanism, crustal rifting and recent mountain building but th
more abundant low temperature fluids are located in most of the world sedimen
tary basins .

The Italian geothermal areas are strictly related to the geodynamic situations
of the territory in respect of the Eurasian plate (Sommaruga ,1980). A young
continental crust is present in the Mediterranean area with collision between
the Eurasian and African plates and related rims divided into microplates and
subject to transport and rotation. This area is characterized by the presence
of cratonic massifs , orogenic systems of the arc-trench type , distensive mar
ginal back-arc basins , volcanic island arcs, active or dormant subduction ar
continental rifting . These geological conditions , with the excetion of sedi-
mentary basins , favour the existence of high enthalpy geothermal anomalies
associated with recent magmatic intrusion and recent or active volcanism .

The Appenines , the bone structure of the Italian peninsula , separate the out
er Adriatic-ionic fore-deep "cold" belt from the inner Tyrrhenian back-arc "Ho
belt .

The "cold" belt , intensively explored for hydrocarbons , shows geothermal
gradients lower than the European average but in favourable conditions inter
mediate temperature aquifers exist at depth in the limestones (and eventuall;
in sands) along with hydrothermal shows at surface (Abano , Acqui and Sirm:
one springs with geothermal fluids above 70°C).

In the "hot" belt, are located the well known Larderello and Amiata dry steam
geothermal fields and the recently discovered water dominant geothermal syst
(Latera in Latium , Mofete in Campania etc). The rift extension rate appar
tly controls the fusion depth and the degree of the mantle fusion in the Tu
cany-Latium geological province (Barberi et al. 1974). Active volcanoes , as
suggested by drilling in the Etna and Vesuvius areas , may show fresh and co:
er than expected waters .

Outside the peninsula , Tertiary and Quaternary rifting in Sardinia is asso
ciated with intermediate temperature springs in the Western areas .

Strictly related with the existing geological conditions are the main Italia
operating or planned geothermal projects for non electric uses , as shown by
table 2 .

In the Po Valley the main utilization is Abano , a world known spa where 130
hotels and 2 ha greenhouses are fed by more than 220 shallow wells .New
wells are under drilling in the Acqui and Sirmione areas where associated di
strict heating and agroindustrial applications might co-exist .

H2

An aquaculture plant in Latisana (Friuli) uses 35°C geothermal fluids from the existing shallow wells . The Rodigo Project(Lombardy) is expected to use the 60°C fresh geothermal waters found at almost 4000 m depth by a dry petroleum wild cat . Another well drilled for petroleum exploration found 100 °C salt waters at 1300 m depth in Mesozoic limestones , to be used for district heating and eventually associated agroindustrial uses in the Ferrara area The Miano project(Emilia) is expected to use the 39°C water of an abandoned petroleum well for biogas production from pigsty waste .

Indirectly connected with geothermal applications are also the ENEL existing experimental plant for thermal agriculture(open field crops with ground heating , sheltered crops with ground and air heating in tunnels and greenhouses) at Tavazzano and the planned pilot plants for fish farming at La Casella and for rice and mais crops irrigation at Pregnana in Northern Italy . In the light of this experimentation mention should be made also of the potential interest of shallow geothermal fluids integrated with effluents from conventional power stations in the Alessandria Lodi belt(Piedmont-Lombardy) for multipurpose applications(balneotherapy, district heatin and agroindustrial uses)asking for an integrated heat pipe-lining system (Lesmo , 1982).

In Central Italy the two main existing utilization poles are Larderello and Mt Amiata . In the Boraciferous Reg on space heating of 1 ha greenhouses is carried out at Castelnuovo by using the effluents from the Larderello geothermoelectric power stations . The under way Pomarance project for greenhouse heating, fish farming , mushroom breeding and small extent electricity production is expected to use the 85-90°C water from an abandoned ENEL geothermal well. In the Mt. Amiata area the under development Pian Castagnaio project uses the 96°C effluents from the local back-pressure local power station ; this project* includes heating of glass greenhouse, drying of agriculture products and district heating totalling 150000-300000 Gcal/y(15000-35000 toe/y equivalent of 60% of the recoverable waste geothermal energy)

In the Latium-Tuscany area , apart from deep high enthalpy geothermal resources for electric uses , intermediate and low temperature fluids are available in the Siena and Radicofan region , around Bolsena and Bracciano as well as between the Colli Albani and the sea . In hhe Siena graben the recently discovered shallow 100°C salt waters could be eventually used for mixed rural(horticulture) and civil(district neating) .Scattered intermediate and low temperature fluids around the geothermal fields or along the outcropping reservoirs could be used at profit by the local rural communities . The Santa Liberata ichtyogenic laboratory of the Orbetello Municipality manages fishing and aquaculture in the local lagoon and uses also salt water from a geothermal well for larvae breeding , prawn reproduction and zooplancton production. The ENEA-ANAPIA farm at Canino is carrying out experiments on geothermal spring waters for greenhouse heating and catfish breeding with computerize systems for heating, farming and roofing research . Finally an important project for a 20 ha greenhouse complex is under development for the Pantani geothermal well near Civitavecchia . Indirectly connected with geothermal activity is also the ENEL thermal aquaculture Valdaliga plant using sea water .

As for Southern Italy a potential utilization pole is represented by the Isch
island , where SAFEN carried out in the past important researches for steam
and developments for new technologies(including an ethyl chloride 250 kW byne
ry cycle power station and a flash pwer plant anticipating foreign experiences and achievements) ; the Ischia geothermal project is expected to use the
thermal wastes from local balneotherapy for integrated cascade applications
(winter heating and summer air conditioning for hotels , space heating for gre
enhouses and fishing farms , servicing for agroindustrial activities)and also
joint fresh water production and electricity generation as in the Vulcano isle
(Sommaruga , 1983).

In the Vulcano island , as favourable permeable conditions are more difficult
to .be found than promising temperature levels, exploration is expected to continue to discover more fluids for joint fresh water production and electricit
generation (CNR, 1981) .

In Sardinia , with intermediate temperature springs(Casteldoria in the North,
Fordongianus   and Sardara in the South) and low temperature hydrothermal
widespread shows, especially in the Western areas , the main aquifer is expec
ted at a 1000 m depth in the Campidano . Because of geographic situations nega
tively affecting other non electric uses , mainly rural geothermal applicatior
are expected south of Decimoputzu(Cagliari) for the existing greenhouses, spe
extension with winter heating of hotels and greehouses in the Sardara area ar
a big close cycle pig growing plant with associated biogas production for the
Macomer—Fordongianus area where the already existing spa is overdimensioned.

As for electricity generation for local rural purposes, apart from pioneering
experiences in Ischia in the period 1942–43, low temperature binary cycle geothermal engines are under develpment for various situations(CNR,1981); a 3
kW engine was installed in 1981 at Abano and a 50 kW engine is going to be installed at Pomarance . Apart from small sized power plants for remote rural
communities , experimentation with satisfactory results is under way for the
Agnano freon geothermoconvector using a downhole heat exchanger to dispense
with the re-injection of geothermal waste water as suggested by the foreign
experiences with single or multiple "U" tubes .

The frequently occurring saline and hypersaline environment of Italian
geothermal fluids(reaching a more than 300000 ppm TDS in Cesano brine and a r
cord value of 515000 mg/l TDS in the Mofete brine) affects negatively both i
direct electric and direct non electric uses , asking for special materials
and technologies in extreme conditions ; as for rural applications , preferab.
asking for fresh warm waters (as in the case of hot irrigation and ground he
ting excluding acidic waters),anomalously heavy brines might prove beneficial
indirectly . In the case of the Cesano brine , the fluid once exploited for
its energy content appears suitable for the production of mixed Na— K fertilizers replacing the presently used K fertilizers as suggested by the Italia
ANIC proposal submitted to EEC in 1979 for recovering the $3K_2SO_4 \cdot Na_2SO_4$ glase
rite and related experimentation under way .

H2.

ECONOMICAL CONSIDERATIONS

The geothermal resources(to be classified for technical and economical purposes accoring to enthalpy, salt content and requested production/utilization technology)include primary thermal fluids (high enthalpy dry/wet steam and pressurized salt waters; medium—low enthalpy salt/sweet waters) from springs, fumaroles and geothermal wells but also secondary thermal fluids (industrial effluents of geothermal and other origin) of equivalent interest.

Notwithstanding abundance(increasing in nature by decreasing temperature), the large scale direct use of medium—low enthalpy geothermal resources is still delayed as compared to indirect electric uses of high enthalpy fluids. Apart from the more profitable geothermal district heating ,no published detail economical analyses refers to the rural geothermal uses reviewed by table 1 as specific temperature ranges . Large investments for well—head production and modified utilization processes coupled with requested high production levels (at least 15000 Gcal/yr for a 1500—2000 m deep well) are the main drawbacks and utilization appears justified only in large scale projects (e.g. greenhouses with a total area of at least 10 ha)(Calabrò and Trombaioli,1984).

In front of transport costs heavily affecting the dispersed geothermal energy high degree—day ratios and consumer densities are required to improve its economics more sensitive to mining risk rather than industry risk and to utilization factor rather than system efficiency; it remains the need to discover the resource near the utilization site and to promote consumer associations . As areal and energetic farm dimensions may be roughly correlated, areal information should be used as a guideline in identifying large and very large farms expected to use industry and energy cost accounting and to have an higher interest in geothermal energy (Lesmo,1985).

As the "hybrid" geothermal project covers aspects of bothsubsurface mining and surface industry , reducing the total prospect risk asks not only for the reduction of the mining risk through the correct choice of the potential resource but also for the reduction of the non mining risk through the correct definition of market oriented rural projects. In matching mining and market , attention should be attached also to the log normal distribution trend of parameters , apperently common to either physical and economical dimensions of geothermal systems as outlined for high enthalpy(Lesmo , 1981) and checked for medium—low enthalpy(Lesmo,1984 and 1985) or areal and economical dimensions of farms (Lesmo,1985); the identfication of distribution trends for basic parameters coupled with the correct choice of probabilistic factors for the project preevaluation and evaluation paralleling resource and reserve assessment may prove a cheap and useful tool in the geothermally oriented territory planning (Lesmo, 1982 and 1985).

TABLE 1 Fluid temperatures ranges requested by agrozootechnic uses, food industries and general servicing.

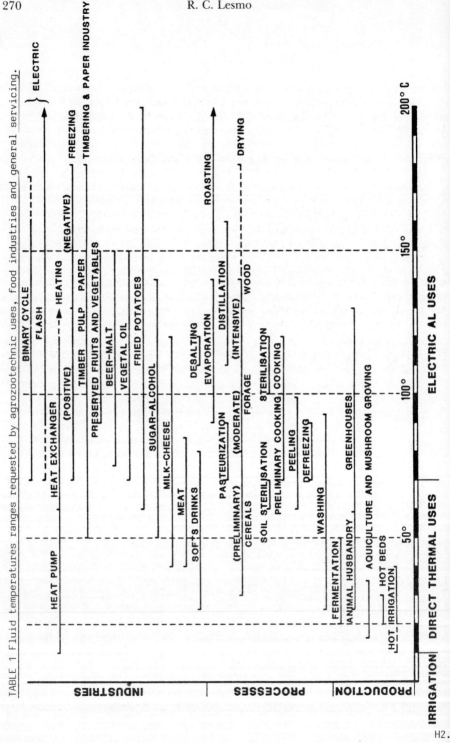

H2.

CONCLUSIONS

Untill recently there was a tendency to think of geothermal development only
in terms of electric power generation , economically more profitable but subject to thermodynamic efficiency restraints by decreasing temperature of the
heat source . The more efficient direct non electric uses , so far often delayed by socio-economical and political situations, affect also agriculture and
related areas, that is animal husbandry, greenhouse heating, hydroponics, aqua-
tic farming , preservation and processing of agricultural products .

The new agriculture philosophy, with emphasis on renewable energy sources and
appropriate technologies for the various conditions, stresses the interest in
using geo-heat by decreasing energy levels not only as a replacement but also
as a complement for both non renewable and renewable energy sources .

The locally available geothermal source,because of consistancy with solar energy and biological cycles, elastic utilization, easy control and maintenance ,
creation of jobs and limited environmental impact , appears promising not only
for rural and island communities but also for other situations thanks to seasonal and cascade integration through heat exchangers and pumps with usual boiler.

Provided sufficient information and assistance are given,the appropriate low
temperature geothermal technology may give a correct answer for a wide range
of situations and the related Italian experimentation, coupled with the sophisticated know-how for high enthalpy problems, may prove useful not only for Mediterranean areas but also for other developing and developed countries .

ACKNOWLEDGEMENT

The author would like to express his gratitude to various CNR, ENEA, ENEL,ENI
geothermal experts, not mentioned or mentioned here only in the text, and especially to the EEC and UN consultant C. Sommaruga for his friendly assistance.

REFERENCES

Lesmo,R.(1981).Criteri e limiti della valutazione delle riserve e risorse geo-
ter.iche.Geologia Tecnica,2
Lesmo,R.(1982).Geologia e programmazione territoriale:problematiche dell'area
termale euganea .Intervento al Convegno di Abano,Dic.1982.Geologi,1 , 1983
Lesmo,R.,and C.Sommaruga(1983).i fluidi geotermici in campo agrozootecnico.
Conferenza Int.CESAT "Energia e Agricoltura",vol.4 Aggiornamenti , Milano
Lesmo,R.(1984).Energia geotermica e geotermia.Giappichelli (Ed.) Torino
Lesmo,R.,and C.Sommaruga(1984). Geothermal fluids in agriculture and animal
husbandry.Proceedings V UN Seminar on Utilization of Geothermal Energy for
Electric Power Production and Space Heating, Florence
Lesmo, R.(1985). Evaluation of geothermal resources and reserves for agricul-
ture.1st FAO Technical Consultation on Geothermal Energy in Agriculture, Rome

TOPIC I
Solar Systems

Solar Technologies

J. A. Duffie

University of Wisconsin-Madison, Madison, WI 53706,
USA

ABSTRACT

In the past two decades, major advances have been made in solar thermal
processes. Important industries have been established supplying equipment,
largely for heating water and buildings. Much of this new technology can
be considered for isolated communities, provided care is used in
identifying the needs, evaluating alternative solutions to meeting them,
and building or buying systems that will perform satisfactorily.

KEYWORDS

Solar energy; heating; cooling; components; systems; performance evaluation.

INTRODUCTION

Solar energy has been under consideration for use in isolated areas for
more than a century. The first large scale application of this energy
resource was in just such a location, on the Atacama desert of Chile where
drinking water was needed for mules used in mining operations. The potable
water was obtained from brines by solar distillation. Operation of the
4000 m^2 still at Las Salinas began in 1872 and continued for more than
thirty five years. The circumstances of that operation were unique. The
site was isolated in the extreme, the mules were needed to work the mine,
and the alternative was bringing in fresh water. The circumstances were
right, the alternatives were much less attractive, and the new solar
technology met the needs.

In the first 70 years of this century, interest in solar energy and its
applications was at a low level. Since 1970, when oil prices started to
climb, there have been major developments in solar energy technologies and
applications (in non-isolated communities). New solar industries have
emerged in several countries, and solar is making significant contributions

to some energy economies. The new technologies have potential applications in isolated communities where conventional energy resources are expensive or not readily available. The status of these processes is briefly noted in the next section.

SOLAR TECHNOLOGIES

In this paper the focus will be on thermal processes for solar energy utilization. Photovoltaic and biological processes may be of great interest in island and rural communities; these will be treated by others.

Solar energy equipment and systems are now being manufactured and sold on a large scale. (In USA, the industry is approaching 10^9/year.) However, while solar equipment is readily available for several applications many processes are in R&D stages, and are not ready for use outside of laboratories or field test situations.

Solar energy systems can be thought of as assemblies of components that provide energy to a load. Quality components and systems have been developed for many applications. It is to be noted that a good system needs more than just good off-the-shelf components. System integration, the assembling of interrelated components into an integrated process, must be carefully done to assure the best possible performance of the process. Examples (in the photovoltaics power field) are provided by Gillett and colleagues (1985) and by Bason (1984), who show that the characteristics of the PV generator, storage batteries, pumps and controls must be matched in order to obtain good system performance.

Components

Many of the flat-plate collectors now being manufactured are of refined design. Durable selective surfaces and low-iron glass, used with modern fabrication techniques, have resulted in better performing collectors than those available a decade ago. Air heaters and liquid heaters are widely available in many designs.

Evacuated tubular (ET) collectors (including some with reflectors to concentrate radiation on the tubes and thus increase their useful output) are now being manufactured. These have loss coefficients that are roughly one-fourth of those of flat plate collectors. They will deliver energy at higher temperatures and will operate at lower solar radiation levels than is possible with flat plate collectors. ET collectors are being used in some commercial water heaters, in industrial process heat experiments, and in absorption cooling R&D.

The problems of design and operation of large collector arrays (with areas of thousands of m^2) are currently being addressed. (See Bankston, 1984.) In contrast to smaller systems, heat losses from piping and control problems arising from dead times and thermal capacitance are often important.

Sensible heat storage in water tanks, pebble beds, or in the structure of buildings remain the most common methods of providing energy storage. We now have better knowledge of how to maintain temperature stratification in water tanks; solar heating systems with liquid heating collectors and stratified water storage tanks can have significant thermal performance advantages over those with unstratified tanks if the collectors are operated at appropriate flow rates. (See Van Koppen and colleagues, 1980, and Wuestling, 1983.)

Many phase change storage systems have come to the marketplace. Their reliability over thousands of cycles has been doubtful, and many of these products are no longer available. When successful, they have the primary advantage of reducing the volume of storage required.

Water Heaters

Solar energy systems for meeting several kinds of energy needs are widely available. The most popular, and probably the most significant worldwide application is solar domestic water heating (DHW). In some countries (e.g., Israel, Australia and the USA) this represents the backbone of the emerging solar energy industry, and a variety of systems is on the market. Common types include forced circulation systems with various freeze protection mechanisms for use in middle and high latitudes, natural circulation systems having their major application in more benign climates, and integral collector storage systems that store hot water in the collector and are used in mild climates.

Space Heating

Space heating is an important energy consideration in temperate and cold climates, and a variety of space heating systems are available. Passive systems have the collection and storage functions integrated into the structure of the building. Active systems have separate collectors and storage, and allow the same levels of building temperature control that can be provided by conventional heating systems. Hybrid systems with active collection and passive storage in the structure itself can meet up to about half of the annual heating loads in temperate climates; these systems are simpler and less costly than active systems, lend themselves to retrofits, work best in masonry buildings, and in some areas are selling well.

Air Conditioning and Refrigeration

It is possible to purchase absorption (heat operated) air conditioning systems designed for solar operation. These units are manufactured in Japan and the USA. They operate with generator temperatures that can be achieved with ET or quality flat-plate collectors. The condensers and absorbers of the coolers must be operated at the lowest possible temperatures. Both of these features are more critical for solar operation than for gas-fired operation. These solar air conditioners are in use in limited numbers for residential and commercial buildings.

Solar refrigeration for food or medicine preservation is still in the experimental state. Most of the experiments are based on the use of intermittent absorption cycles, for making ice or for chilling a small box. There has also been work on photovoltaic powered mechanical refrigerators. This equipment may be approaching the point where it can be built or bought with confidence that it can operate satisfactorily over many years.

Drying

Many experiments have been done on solar drying of timber, grains, fruit
and other agricultural products. This topic will be discussed by others at
ERIC IV. No clear pattern is emerging from these experiments. The most
significant results may be the determination of useful drying cycles and
specifications for dryer and collector construction. For most solar drying
processes, air temperatures needed are not much above ambient temperatures,
and collector design is not as critical as with higher temperature
processes. On-site construction of solar dryers, if well done, may be
feasible (in contrast to a process like air conditioning, where factory
built equipment will be the only alternative).

Performance Estimation

A rational evaluation of an energy system must include estimation of annual
energy delivery. This is not always easy, particularly as performance is
dependent on how well the system is designed and constructed, and on the
weather and loads to which it is subjected. In contrast to many
conventional energy sources where system overdesign costs little, the
nature of costs of solar energy systems (high capital costs and low
operating costs) mean that overdesign is very costly. Thus sizing a solar
energy system is critical, and estimation of thermal performance is an
important part of the design and evaluation process.

As space and water heating systems became more widespread, easy-to-use
methods of estimating their thermal performance were developed. The most
widely accepted methods in North America are included in the FCHART
programs which use monthly average meteorological and load data to estimate
annual performance. The algorithms in FCHART are a combination of
correlation methods (Beckman and colleagues, 1977) and utilizability
methods (Klein and Beckman, 1984). Systems that can be designed with
FCHART include DHW, space heating with active systems utilizing water tank
or pebble bed storage, industrial process heating, passive heating by
direct gain or collector-storage walls, hybrid systems with active
collection and passive storage, other hybrid systems, and swimming pool
heating. Many photovoltaic processes can also be evaluated by closely
related methods that are based on the same utilizability concepts as are
used in designing passive heating systems.

Other design methods have been specifically developed based on European
meterological data (Kenna, 1984) and additional methods for active and
passive systems are in preparation by the EEC. Many passive building
heating systems in USA are designed using the Solar Load Ratio method
(Balcomb and colleagues, 1984).

For systems that do not fit the standard configurations for which these
design methods were developed, it may be necessary to resort to simulations
to evaluate performance. Simulations are hour-by-hour calculations of how
systems perform. They are based on hourly meteorological data, and are
simultaneous solutions of the sets of equations that describe the
performance of the components. Simulations are more difficult to use than
design methods and require much more detailed weather data, but they
provide information on the dynamics of systems operation (i.e., how
temperatures and energy transfer rates vary, and what extremes are to be
expected) as well as on the annual performance. There are few practical
limits on the types and complexities of the systems that can be

simulated.   For example, combinations of solar and other energy resources
such as wind can be handled if the characteristics of the components are
known   and   the   needed   meteorological   data   are   available.   Simulation
programs can be modular, general purpose programs like TRNSYS (see Klein
and colleagues, 1984), or they can be special purpose programs that are
useful only for specific types of systems such as that used by Bason (1984).

There are practical limits to predicting performance.   Predictions must be
based on historic (i.e., average) weather data, which may not be very good,
and there is no way to predict what weather and thus performance will be in
a particular year.   Also, all of the prediction methods are based on the
assumption that systems will be well built.   Many systems do not perform up
to expectations because they are poorly constructed or installed.

Economic and Social Factors

Given estimates of average thermal performance, the question remains of
estimation of economic feasibility.   There is no universally accepted
method or criterion for making comparisons.   The economic problem is one of
comparing the costs of investments with reduction of future operating
costs.   The method built into the FCHART and TRNSYS programs is a life
cycle cost method that is used by the government and some (but not all)
solar industries in the USA.   The results of LCC analyses are dependent on
the assumptions made about future costs of fuel, interest and discount
rates,   period   of   the   analysis,   possible   resale   value   of   the   solar
equipment, and other economic parameters.

Factors other than economics can be of critical importance in evaluating
and selecting energy resources.   These can include reliability, aesthetic
values, conservation of resources, independence of outside sources, etc.,
some of which may be quantifiable to some degree.   There may be "people
problems" that will have strong influence on energy decisions.   These can
include   adaptability   to   new   technologies,   education   requirements,   the
realities of energy needs, people's perceptions of a new technology, peer
practices, and other social factors (Miller and Duffie, 1970).

ASSESSING ENERGY APPLICATIONS

Solar processes are at a state of development where they can be considered
for meeting a spectrum of energy needs.   Inevitably there is range of
sources   (solar   and   other)   and   processes   that   can   be   considered   as
candidates for a particular application.   The following are suggested
considerations in assessing and selecting energy processes in isolated
communities.

Assessment of need is a first step, and should include consideration of the
type of energy needed (thermal, electrical, etc), the time span of needs
(seasonal, night, temporary, etc.), temperatures at which energy is needed,
and other similar factors that affect energy use.   There have been examples
of misjudgment of energy needs; an objective of a careful assessment of the
situation is to avoid an expensive solution to a non-problem.

Alternative   solutions   to   energy   problems   are   always   available.   For
electrical energy needs, solar photovoltaics, diesel generation, wind,
connection to power grids however remote, and others may be considered.
From the users point of view, the problem should be thought of (for

example) as supplying water and not as building a solar still. Many of the alternative solutions can be quickly ruled out on the basis of cost, state of development, climate or other factors. The nature of costs of processes may be important; finding funds to invest in capital-intensive energy systems may be a major problem in communities that have limited capital resources, no matter how much operating expense is saved.

Assessment of social factors is important, particularly in isolated communities with developing economies and particularly for applications on a household scale where problems of educating individual users about a technology may be critical. It may be much easier to introduce a technology on a community or small industrial scale, where relatively few operators will need to be trained.

Assurance is needed that the system proposed will be based on proven technologies and that they will deliver the expected energy or product.

For solar applications, estimation of annual energy contributions can be done by simulations or by design methods. Other candidate energy systems should be subject to the same quantitative evaluation.

Only when this range of problem areas has been fully explored can rational decisions on selection and application of energy in rural and island communities be made. Based on these studies there will be many energy problems that will be best solved by solar energy.

REFERENCES

Balcomb, J.D., and colleagues, (1984). Passive solar heating analysis, ASHRAE, Atlanta.

Bankston, C.A., and colleagues, (June, 1984). Design and performance of large solar thermal collector arrays, Proc. of IEA Workshop, San Diego, Report SERI/SP-271-2664.

Bason, F., (1984). Performance of a small photovoltaic power supply with Ni-Cd storage. Proc. ERIC III, Pergamon Press, Oxford.

Beckman, W.A., Klein, S.A. and Duffie, J.A.,(1977). Solar Heating Design by the f-Chart Method, Wiley, New York.

FCHART-4 User's Manual, Version 4.2, (1985). EES Report 50, University of Wisconsin-Madison.

Gillett, W.B., Derrick, A., and Kropacsy, C.C.J., Test methods for photovoltaic water pumping systems. Presented at INTERSOL 85, ISES Congress, Montreal.

Kenna, J.P., (1984), A parametric study of open loop solar heating systems-I; A parametric study of open loop solar heating systems-II. Solar Energy, 32, 687 and 707.

Klein, S.A. and colleagues, (1984). TRNSYS, A Transient System Simulation User's Manual, University of Wisconsin-Madison, EES Report 38-12.

Klein, S.A., and Beckman, W.A., (1984). Review of solar radiation utilizability, Solar Energy Engineering, 106, 393.

Miller, R.J., and Duffie, J.A. (1970). Thoughts on economic-social implications of solar energy use. ISES, Melbourne.

VonKoppen, C.W.J., Thomas, J.P.S., and Veltkamp, W.B., (1980). The actual benefits of thermally stratified storage in a small and a medium size solar system, Eindhoven University of Technology, The Netherlands.

Wuestling, M.D., M.S. Thesis, (1983). Investigation of promising control alternatives for solar water heating systems. Mechanical Engineering, University of Wisconsin-Madison.

Solar Steam Generating System for Agro Based Industries

A. Thomas*, M. Ramakrishna Rao*, J. Nagaraju*, H. M. Guven** and R. B. Bannerot***

*Instrumentation and Services Unit, Indian Institute of
Science, Bangalore 560 012, India
**Mechanical Engineering Department, San Diego State
University
***University of Houston, USA

ABSTRACT

A solar steam generating system for a silk industry in India has been conceived.
The design aspects of the system, the development of parabolic trough concentrators (PTC), the optical and thermal evaluation of PTC are presented in this
paper.

KEY WORDS

Sericulture; flash boiler; optical efficiency; intercept factor;economics.

INTRODUCTION

Agro based industries play a very important role in the economy of the developing
countries like India. The availability of cost effective, reliable and safe energy
sources would raise the standard of living of the rural community. A programme
has been undertaken to design, develop and install a solar steam generating
plant at the Government Silk Weaving Factory, Mysore, India.

Sericulture is one of the major rural based agro industries of Karnataka and
the state produces good quality silk for internal consumption and export. The
solar steam generating plant under construction will supply low pressure steam
to the printing section of the silk factory. The printing process consists of
warming of the fabrics laid on a well prepared table heated to a temperature
of 45°C and using screen printing techniques. The heat is supplied by passing
the steam through the pipes laid underneath the printing tables. After air
drying, the colours on the fabrics are fixed in an ager, heated by steam. The
used steam from the printing table will be used to heat water which in turn
will be used for washing the printed fabrics. Steam is also used for other
purposes like screen drying, gum cooking and colour mixing. The estimated
requirement of steam is 250 Kg/hr at a temperature of 150°C and at a pressure
of about 4.56 bar. At present, this requirement is met from a coal fired locomotive
boiler. In this project steam will be supplied at the rate of 90 Kg/hr only
with the collector field having an area of 180 m^2.

DEVELOPMENT OF PARABOLIC TROUGH CONCENTRATORS (PTC)

The concentrator field forms about fifty percent of the cost of the solar steam generating system. In order to make the entire system cost effective the thermal efficiency of the concentrator modules must be of the order of 60 percent at a temperature of 250°C. Efficient performance of PTC demands stringent specifications on the reflector support structure, reflecting surface, receiver tube coating and tracking sub system. Development of concentrator modules for high reliability and low cost is a continuous evolution process and hence three types of reflector support structures were conceived of viz., semi-monocoque, sandwich and stiffened rib structure. These structures were developed in order to meet the various design goals like

* ability to maintain its optical and thermal performance during normal wind speed (48 Km/hr and survive the extreme wind speed (144 Km/hr) conditions).
* to have high stiffness/weight ratio
* to have low cost/effectiveness ratio
* amenable for mass production techniques
* ease of transportation, installation and maintenance.

The construction of semi-monocoque and sandwich structure involves highly labour-oriented production techniques which may limit the quality and precision of the final product and efficiencies. Hence a stiffened rib structure (Fig.1) having an aperture area of 6.4 m^2 was developed involving advanced production techniques. Detailed studies conducted on the above structure satisfies most of the design goals indicated above and hence this concept was adopted for the development of the solar collector field required for the solar steam generating system. Table I gives the specifications of the PTC adopted for solar steam generating systems.

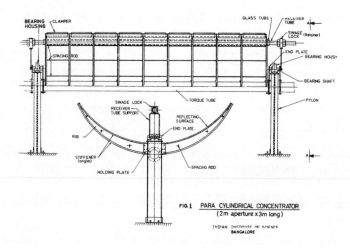

Fig. 1. Para cylindrical concentrator.

TABLE I   Specification of PTC

Aperture area	: 6.4 m^2 (2.1m x 3.05m)
Focal length	: 0.506 m
Rim angle	: 92°
Receiver tube - O.D.	: 33 mm
I.D	: 28 mm
Glass envelope O.D.	: 59 mm
I.D.	: 54 mm
Reflectivity of mirror	: 92 percent
Absorptivity of the receiver	: 0.94
Emissivity of the receiver	: 0.15

OPTICAL PERFORMANCE EVALUATION OF PTC

The instantaneous efficiency of PTC is largely determined by its optical efficiency (η_o) and thermal losses.   For a given concentrator, the optical efficiency is governed by its intercept factor (γ) which is defined as the fraction of rays incident upon the aperture that reach the receiver for a given incidence angle. The variation of (η_o) with (γ) is shown in Fig.2.   The intercept factor is the parameter that embodies the effect of errors.   The errors that occur in PTC are well presented by Guven (1983) and shown in Fig.3.   Out of these, the local slope and profile errors occur during manufacture.   Thomas (1986) has developed a technique to measure the flux distribution around the receiver of PTC.   The theoretical and the experimental distributions for a prototype with stiffened rib structure are shown in Figs.4 and 5.   Knowing γ one could determine the manufacturing errors, σ and thus η_o.

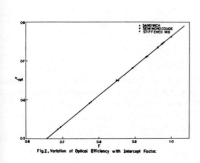

Fig.2. Variation of Optical Efficiency with Intercept Factor.

FIG.3 DESCRIPTION OF ERRORS

Fig. 2.   Variation of optical efficiency   Fig. 3.   Description of errors.
with intercept factor.

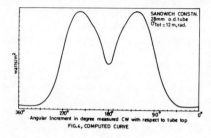
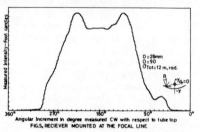

Fig. 4. Computed curve.   Fig. 5. Receiver mounted at the focal
                                              length.

Table II   The manufacturing error (σ) intercept factor (γ) and
           the optical efficiency (η_o) of three concentrators

	σ	γ	η_o
Semi-monocoque	16.3	0.83	0.63
Sandwich	23.2	0.89	0.68
Stiffened rib	12.7	0.92	0.704

THERMAL PERFORMANCE EVALUATION OF THE CONCENTRATOR MODULES

Based on the ASHRAE a standard test facility for PTC has been established.
The instantaneous efficiencies of the three models have been determined and
are shown in Fig.6. The performance of the PTC with stiffened rib structure
is much better than the other two modules and is comparable with the acurex
collector tested by Sandia Laboratory, USA.

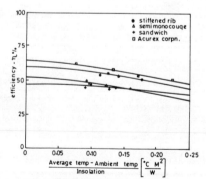

FIG 6 PERFORMANCE OF THE THREE MODELS OF THE
       CONCENTRATOR.

Fig. 6. Performance of the three models
        of the concentrator.

SYSTEM DESCRIPTION

As shown in Fig.7, the system consists of a distributed field of parabolic trough concentrators, a flash boiler, boiler feed pumps, circulating pumps, valves, piping instrumentation and controls. In operation, heated water at 35°C is fed from the water treatment plant through a feed pump. The flow of water into the flash boiler at 7.09 bar is controlled by an electrically operated solenoid valve which opens or closes based on the water level in the flash boiler. The water enters the flash boiler through a deaerating heating arrangement located on top of the flash boiler drum. In this arrangement an intimate mixing of the incoming water is effected with a portion of the steam generated in the flash boiler.

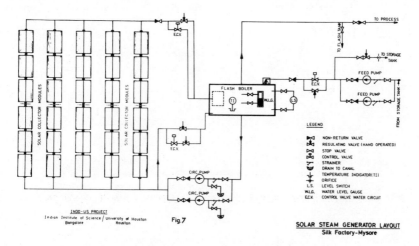

Fig. 7. Solar steam generator layout.

The water thus heated enters the solar collector modules through a circulating pump which pressurizes water at 10.73 bar. The water is heated to 175°C and is fed into the flash boiler. In the flash vessel water at 10.73 bar and 175°C is flashed to steam at 4.56 bar and 150°C. The flashed steam is taken out from the flash boiler through a single outlet connected to the existing steam mains in the plant.

The system is provided with suitable instrumentation and controls to maintain water level in the flash boiler, pressure and temperature in the solar collector outline and pressure in main steam line. The solar steam generating system was designed based on the average solar insolation level at Mysore, steam require-ment, concentrator efficiency and the heat balance of the system. The system specifications are given in Table III.

ECONOMICS

An economic   analysis of the solar steam generating system has been performed using the standard technique described by Dickinson (1977). The analysis is based on the actual cost of the system, interest rates prevailing in India and the actual consumption of coal. The pay back period for the system is around 15 years considering escalation of the coal at 10 percent. The pay back period will be reduced to about 10 years, if large systems are installed and mass produc-tion techniques are adopted for the manufacture of concentrators.

TABLE III System Specification

Collector field

Type	:	Line focus parabolic trough
Area	:	180 m^2
Orientation	:	North-South
Tracking	:	With flux line sensors
Heat transfer medium	:	Water

Flash Boiler

Capacity	:	1.26 m^3
Design pressure	:	14.6 bar
Design temperature	:	180°C
Flow rate	:	2090.4 Kg/hr

CONCLUSION

The solar steam generating plant which is being installed is the first of its kind in India. The experiences gained in the application of the solar energy in silk manufacturing industry will not only supplement the conventional energies but also indirectly helpthe industries in the rural areas in the long run.

ACKNOWLEDGEMENTS

The authors are grateful to Department of Non-Conventional Energy Sources, Government of India and USAID for financial support. Our thanks are also to Mr.G.K.Muralidhar in assisting us in the experiments.

REFERENCES

Dickinson, W.C. and H.J. Freeman (1877). Lawrence Livermore Laboratory, VCRL - 52254.
Guvan, H.M. (1983). A transportable Fortran based system for computer design of parabolic cylindrical solar energy concentrators. Ph.D. dissertation, Mechanical Engineering Department, University of Houston, Texas, USA.
Thomas A., Ramakrishna Rao, M., Guven, H.M., Balasubramanian, V. and Shankar, G. (1986). Communicated to ASME Solar Energy Division Conference at Anaheim, CA, USA.

Simple Solar Systems for Water Heating in Gran Canaria

P. Kjaerboe

VIAK AB, Consulting Engineers, Box 519,
S–162 15 Vällingby, Sweden

ABSTRACT

At a physiotherapeutic treatment unit, four solar heating systems were built between 1982 and 1983. The total solar collector area is 460 m^2, 4600 ft^2.

During the design period, 20 other systems, with the total area of 2600 m^2, 26 000 ft^2, were inventoried. Special problems, such as corrosion from salt groundwater and saline environment, were noted. Furthermore, cover material had been degraded because of solar radiation and insulation was often ignored.

Some improvements have been made of the systems and suitable materials have given recently built units a satisfactory lifetime of more than ten years. Furthermore, installation costs and savings are such that the energy price of water within the range of temperatures 20° - 50°C, 70° - 115°F, is in the same range as the price of fuel oil, i e without any support and with a calculated interest of 10% and a length of service of 10 years.

KEYWORDS

Solar heating, water heating, survey, Canary Islands, economy.

BACKGROUND

One of these islands, Gran Canaria, picture 1, is permanently inhabited by 820 000. About one fourth as many or an annual total of 200 000 tourists visit the island with arrivals and departures every other week, some staying for 8 or 9 months. This creates increased demand on service of all kinds.

The price of energy is high, 0.6 to 1.0 SEK/kWh or 7-12 cents/kWh for locally generated energy is common. On the mainland the consumers' price is reduced. Consumers' price is then between 0.25 (gas), 3-6 cents/kWh, and 0.5 (electricity) SEK/kWh or 6 cents/kWh.

I3.1

288   P. Kjaerboe

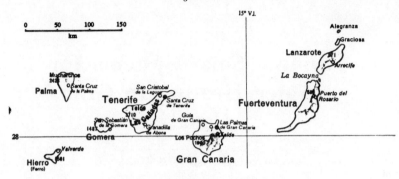

Picture 1   Canary Islands, Spain, are situated on
             lat 28°N, long 15°W.

PLANT

In a medical care centre, owned by the Swedish Employers Federation, where
there are heated pools for water exercise, a survey was made with a view
to reducing cost of heating.

The pool temperatures were 32°C and 24°C, respectively.

DOCUMENTATION

A documentation of older solar heating systems was made in order to learn
from experiences gained. Totally 23 plants with a collector area of 2260 m^2
were studied. Design data such as solar flux, altitudes, ambient and well
water temperatures were found in handbooks published by Spanish authorities.
The sunshine hours exceed 2800 h and the flux varies between 4.0 and
4.9 kWh/m^2 day depending on the month.   Metering of temperatures and collection
of fuel consumption data were also carried out as a design basis. The energy
demand in general for DHW varies between 3 and 4 kWh/person and day.

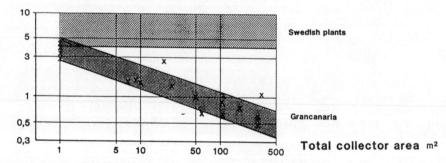

Picture 2   The chosen collector-surface area was de-
             creasing when the size of the plant was
             larger. This is not the case in Swedish
             plants.

I3.2

MEASURES

As the first measure, a separation of the two systems was made between the
two pools heated. A large amount of energy was then saved as the losses
depend on temperature, and increase exponentially with a factor larger than
one. To decrease losses during nights, covers were also suggested – a roll-
able and stiff type, respectively. The expected saving with covers installed is 30 % depending on how often they are used.

SOLAR COLLECTORS

As the next step to reducing energy consumption, solar collectors were suggested. Some of the claims on such a system were:

- easy to build with available tools, and easy to repair

- spare parts should be available.

Also as elementary claims the following were put forward:

- temperatures of $24^{\circ}C$ and $32^{\circ}C$ should be reached

- the material chosen should resist the environmental influence, both from
  water and the atmosphere containing both salt from the Atlantic and sand
  dust from the Sirocco.

Collectors evaluated from the points of view mentioned above were one glazed (single glass) and one uncovered. The uncovered one, picture 3, consists
of coils of 110 mm or 4.4 in. width delivered in lengths of 180 m or 600 ft.

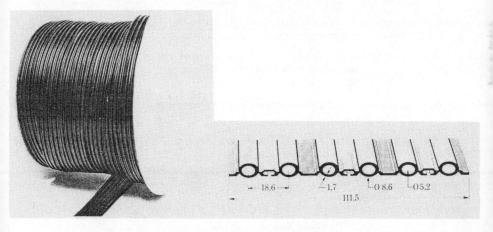

Picture 3   The uncovered collector made from EPDM-rubber
is delivered in coils. N.B. different scale.

The length of the collector in the installation can then be freely chosen.
Connections with main pipes are easily made with teflon jamb sleeves, see
picture 4. The material of this collector is extruded EPDM-rubber a UV-
stabilizer as component.

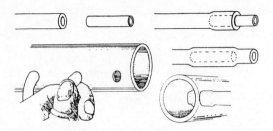

Picture 4   Connections are made to the main supply heat-
             er with a jamb sleeve made of teflon.

Apart from the glass cover,the glazed collector, picture 5, also consists
of absorber, insulation and envelope. Connections to the collector are
made at two ends with couplings which are soldered.

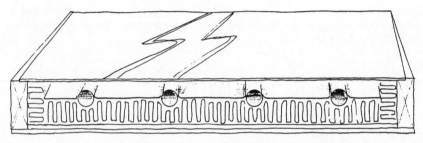

Picture 5   The insulated and covered collector has an
             other unit size.

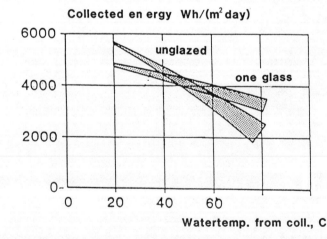

Picture 6   Efficiency as a function of temperature out
             from the collector.

EFFICIENCY

The two collectors were compared regarding efficiency and temperatures needed. Picture **6** a chosen November day, shows the amount of energy collected as a function of temperature. In the prevailing temperature intervals the uncovered collector seems to have advantages.

MATERIAL

The water in the pools has rather high contents of chlorides and one has to consider using either heat exchanger or resistant material. The EPDM-rubber is then favourable.

Sand in the form of dust will be collected on the surface, especially after the Sirocco. A surface of glass will easily be cleaned.

SYSTEM

From Holmberg (1979) system costs were mentioned. A surprisingly large part of the cost is due to installation and work (see picture 7). The two collectors have different advantages, for example the number of connections.

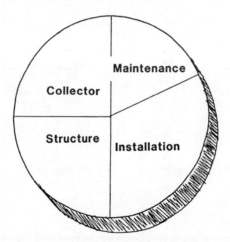

Picture 7   The costs for an installation can be divided
              into running and maintenance, installation,
              structure and collector.

Totally eight connections with the unglazed collector were necessary for a 145 m^2 plant. Compare this to the fact that the glazed unit would normally need approximately 200 connections (see picture 8).

The uncovered rubber collector was finally chosen.

The stand was made in situ from known material and with an optional shape depending on the accessibility (picture 9).

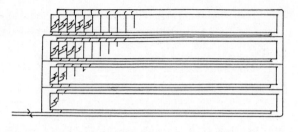

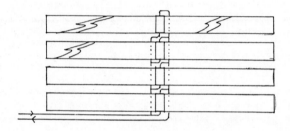

Picture 8   The number of connections can be reduced when
             choosing a system.

Picture 9   In this case there was a limited option for
             location of the solar collector.

If spare parts should be delivered toghether with the plant itself it is
suggested that the amount is kept as low as possible. When this type of
collector is chosen, the list of spare parts is restricted to the mat.

CONCLUSIONS

The survey and experiences from the built plants show that:

- systems designed and built by others than the local inhabitants do not use locally used and known technique or materials,(example see picture 10).

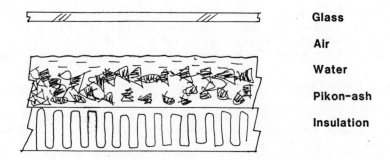

Glass

Air

Water

Pikon-ash

Insulation

Picture 10   Example of a solar collector made of vulca-
            nic ash or pikon.

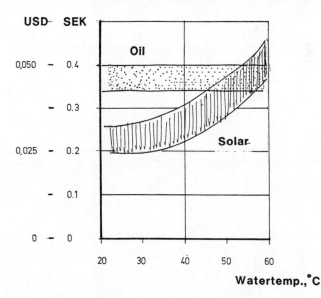

Picture 11   The cost of heat in this case was in the
            same range as that for oil.

- running maintenance and service has to be thoroughly understood when introducing new installations and materials

- the circumstances are exceptional, e.g. extremely high insolation and
hard environment causing corrosion, and high prices of conventional and
competing energy

- other plants on the Canary Islands are more expensive compared to plants
built in Sweden.

- the cost for energy in this case has been evaluated and with an expected
lifetime of 10 years and an interest rate of 10% the price is in the same
range as that of oil,(see picture 11).

REFERENCES

Kjaerboe, P. and Öst, S.: Enkla solfångare för varmvattenvärmning, del 1.
Några problem och resultat från Gran Canaria TM, Institute of Technology,
Stockholm, Dep of Heating and Ventilation 1985.
(Simple type solar collectors for domestic hot water heating, part 1.
Some problems and results from Gran Canaria)

Peterson, F. and Sandesten, S.: Solar heated domestic hot water. Swedish
Council for Building Research, R145:1984.

Holmberg ,I. and Kjaerboe, P.: Solar heated municipal swimming pools,
metering 1978. Swedish Council for Building Research, R39:1979.

The Role of Electric Utilities in Promoting Solar Heating in Rural and Island Communities

K. MacGregor

Napier College, Colinton Road, Edinburgh, UK

ABSTRACT

In this paper the case for and against the involvement of electric
utilities in solar heating is analysed and methods of involvement are
reviewed. A case study is presented of a proposal for utility-assisted
solar heating in the Shetland Islands of Scotland. The proposal appears
economically attractive for both the utility and its customers. It is
concluded that there is generally a good case for electric utility
involvement in solar heating, particularly in rural and island areas.

KEYWORDS

Solar heating; electrical utilities; rural and island communities.

INTRODUCTION

Unlike the renewable energy sources (such as wind, wave, hydro and
photovoltaics) whose output is electricity, there is a less obvious
interaction between solar heating and the electricity producing utilities.
However, if solar heating of water or buildings displaces electricity then
it will affect both the pattern and the total amount of electrical demand
and it is desirable therefore that electric utilities should be aware of
the technical and economic implications of solar heating and should
formulate appropriate policies towards it.

MOTIVATION FOR UTILITY INVOLVEMENT

The case for the promotion of solar heating by a particular utility depends on:its existing supply infrastructure; the present and predicted future pattern of demand; the existence of physical, political or economic constraints on further development of the supply infrastructure;and the presence of legal/regulatory limitations or directives concerning its involvement with the utilisation rather than merely the supply of electricity.

Six main positive motivations for utility involvement have been identified:
1. <u>Reducing the need for additional generation/transmission capacity</u>.
   This applies specially to utilities with a growing summer-peaking demand. Solar water heating which yields maximum savings in summer and also provides a degree of diurnal storage can reduce the peak demands which are likely to occur during the daytime on hot sunny summer days when airconditioning and refrigeration plants are operating at maximum capacity. There may also be a case for solar space heating in areas with a winter peak demand since there appears to be a correlation, albeit a weak one, between the incidence of the coldest and sunniest days. In addition, several passive solar heating techniques such as retrofit sunspaces and solar skins can increase the thermal resistance of the external fabric of a building and can therefore reduce peak heating loads regardless of the availability of solar radiation.

2. <u>Reducing Sales of Electricity</u>. If electricity is being sold at a marginal loss due to tariff anomalies or legal/political constraints then solar heating of both space and water will reduce sales from and losses to the utility.

3. <u>Increasing Sales of Electricity</u>. For utilities which have over-invested in supply capacity and who therefore wish to increase sales the concept of the solar + electric heating combination may be an attractive marketing strategy. This has already proved to be the case with the combination of insulation + electricity in new housing in the UK where it is claimed that off-peak (or "white-meter") electric heating is now competitive with mains gas in well insulated houses. In purely economic terms insulation is usually more cost-effective than solar heating. On the other hand, solar energy can make a positive contribution to the water heating load rather than merely reducing it while certain passive solar heating approaches such as the attached sunspace or conservatory uniquely provide valuable benefits of improved amenity and comfort, even though these may be difficult to quantify, in addition to energy saving.

   The flexibility of electricity as a heating medium generally makes it well suited as a back-up to solar heating. The provision of off-peak hot water storage capacity (for both space and domestic water heating) can be shared by active solar heating systems.

4. <u>Diversification</u>. In addition to the supply of electricity many utilities have already diversified into the marketing, installation and maintenance of electrical appliances. Incentives for doing so include the desire to sell more electricity and the ability to set standards of safety and efficiency. There may be a case for extending this to solar heating equipment which is designed to be complementary to the electrical supply.

5. <u>Compulsion</u>. There may be directives from governments or regulatory
   bodies for utilities to become involved with renewable energy sources
   such as solar heating.

6. <u>Information</u>. Even though an electric utility does not intend to become
   immediately involved in solar heating it may still decide that, in
   order to be fully informed about the present state of the art and its
   future potential, it should support research, development and trials
   of solar equipment.

METHODS OF INVOLVEMENT

In order of increasing involvement, the options appear to be:

1. <u>Provision of General Advice to Consumers</u>. This assumes that the
   utility has access to sources of information on solar heating or has
   conducted its own investigation programme.

2. <u>Energy Surveys and Advice to Individual Consumers</u>. Solar heating could
   be included among the options recommended to individual householders
   following an energy survey conducted by or on behalf of the electrical
   utility. Advice could be given on design and sizing of systems with
   guidelines on the savings which might be expected.

3. <u>Supply, Installation and Maintenance of Solar Heating Systems</u>. This
   would be done on the same basis as for electrical appliances and could
   be seen as an extension of existing practice. A large customer such as
   an electric utility should be able to set standards and specifications
   for equipment and even to encourage, develop and test innovative designs.

4. <u>Provision of Financial Incentives</u>. These could include low-interest
   loans, grants, rebates or re-structuring of tariffs. The degree of
   financial assistance which a utility might justifiably offer would
   obviously depend on the savings in capital and operating costs which
   it would expect to obtain as a result of the installation of solar
   heating equipment or systems by its customers.

JUSTIFICATION FOR NON-INVOLVEMENT

Arguments put forward both by utilities and others against involvement by
utilities include:

1. <u>Adverse Effect on Load Factors</u>. If many consumers instal solar heating
   this can reduce load factors, specially for winter-peaking utilities.
   The utility may still have to provide plant to meet the same peak
   demand in cold, cloudy weather but will lose sales during the summer.
   There may also be quite large short-term variations in load depending
   on the day to day pattern of weather. On the other hand, the in-built
   storage capacity of most solar heating systems can help to smooth the
   short-term peak demand on the whole electrical system.

2. <u>Market Distortion</u>. It has been argued that direct investment or other
   involvement by large utilities can distort the market, lead to unfair
   or monopolistic trading practices and stifle innovation. It has even
   been alleged (Bossong, 1982) that utilities might enter the solar market
   merely for public relations purposes or to discredit solar technologies.

3. Inequity. Not all consumers can benefit equally from utility assisted
    solar heating. Those who already use above-average amounts of
    electricity and whose houses are well sited could benefit at the expense
    of other consumers who are more thrifty, poorer, less well situated or
    who have already installed solar heating at their own expense. On the
    other hand, if a programme of utility investment in solar heating leads
    to overall savings to the utility, then all the consumers should benefit
    to some extent.

CASE FOR UTILITY ASSISTED SOLAR HEATING IN SHETLAND

Electricity is generated and supplied in the Shetland Islands of Scotland
by the North of Scotland Hydro Electric Board, a public utility. They
supply 8860 domestic consumers with an approximate average annual
consumption of 7600 kWh per consumer. Generation is from 60 MW installed
capacity of diesel sets. The maximum demand on the system is approximately
30 MW. Due to political constraints the utility is obliged to offer the
same tariffs on Shetland as they do on the mainland of Scotland where
generation is based on cheaper coal, nuclear and hydro sources. Consequently
each unit is sold at a loss, even at the margin. A reduction in demand,
achieved by any means, will reduce these losses. In this section, the extent
to which the Hydro Board might be justified in assisting solar heating for
its domestic consumers in order to reduce its losses will be examined.

Marginal Losses. The marginal cost of supplying electricity in Shetland is
5.05 p/kWh (North of Scotland Hydro Electric Board, 1985). About 95% of
this is due to fuel costs. The marginal loss to the Board is shown in Table 1

Tariff (Domestic Consumers)	Selling Price p/kWh	Marginal Loss p/kWh
Standard Tariff	4.65	0.40
White Meter Economy Tariff, Day Rate	4.91	0.14
White Meter Economy Tariff, Night Rate	2.02	3.03

TABLE 1 - Marginal Losses of Electricity Supply in Shetland

Solar Energy Savings. The annual energy savings resulting from four solar
heating options installed in domestic buildings in Shetland have been
estimated and are shown in Table 2.

System	Capital Cost (£)	Annual Energy Saving (kWh)	Ratio (kWh/£)	Annual Savings Capital Cost
Solar Water Heating (4 m²)	1000	1380		1.38
Attached Sunspace (16 m² wall cover)	700	1764		2.52
Solar Skin (20 m² wall cover)	500	2800		5.60
Sunspace + Water Heating (16 m² + 4 m²)	900	2575		2.86

TABLE 2 - Estimates of Costs and Savings for Four Solar Options

The estimate of savings for the solar water heating system is based on
British Standards BS 5918 (1980) modified for location in Shetland
according to Reynell (1984). The sunspace savings are based on experimental
work by Ford and Everett (1985) modified by estimates by Bartholomew (1984)
of the relative performance of sunspaces in the extreme North and South of
the British Isles. Solar skin is a technique for external cladding of
masonry buildings using translucent materials. Estimates of savings are
based on work by MacGregor (1979). The estimate of savings from the
combination of a sunspace with integral solar water heating is based on
unpublished work by MacGregor.

Cost Savings
Assuming (i) that the above energy savings are achieved in domestic
dwellings with electrical heating of water and/or space, (ii) that
electricity is supplied under the White Meter Economy Tariff (off-peak)
with two-thirds at the night rate and one-third at the day rate and (iii)
that the Hydro Board were to provide 50% of the capital cost of the solar
heating equipment, then the rate of return on investment for the Hydro
Board and the simple pay-back time for the consumer, who would pay the
remaining 50%, can be estimated and are shown in Table 3.

System	Annual Rate of Return for Hydro Board	Simple Payback Time for Consumer
Solar Water Heating	5.7%	12.5 years
Sunspace	10.4%	6.8 years
Solar Skin	23.2%	3.1 years
Sunspace + Solar Water Heating	11.8%	6.0 years

TABLE 3   Economic Returns for Utility-Assisted Solar Heating

In all cases the estimated rate of return on investment for the Hydro Board
is above the minimum (5%) set for nationalised industries in the UK. For
consumers, the pay-back times in most cases are sufficiently attractive to
justify taking out a loan at commercial rates for their share of the capital
investment. In addition, the two sunspace options will provide a bonus of
improved amenity and space as well as energy and cost savings.

Direct investment in energy saving by a utility such as the North of
Scotland Hydro Electric Board may seem a radical departure for an
organisation which has traditionally perceived its main function as merely
the supply of electricity. However the Act which set the Board up in 1943
urged the new body to "collaborate in carrying out any measures for the
economic development and social improvement in their area, so far as their
powers and duties permit". In addition, the recently revised statement of
objectives for the Scottish Electricity Boards (1983) includes the
directives:
(a) "In developing capital investment the Boards' objective should be the
    supply of electricity at the lowest possible cost consistent with the
    required rate of return on new investment".
(b) "The North of Scotland Hydro Electric Board should take all practicable
    steps to reduce the cost of diesel generation".

(c) "The Boards should encourage the conservation of electrical energy and
    the development of renewable sources".

The above proposal for direct investment by the Hydro Board in solar heating
in Shetland appears to meet all these criteria and objectives.

CONCLUSION

In general, there appears to be a sound case for involvement by electrical
utilities in the promotion of solar heating in order to reduce their running
costs or as an alternative to investment in additional supply capacity.
The risks of investment in a large number of small scale energy saving
installations are less than for a single large generation plant. Methods of
involvement range from the provision of general advice to the provision of
financial incentives and the design, supply and installation of solar
heating equipment. The case for utility assisted solar heating (and other
energy saving measures) is particularly strong in rural and island areas,
many of which are served by diesel generators with high running costs.
Solar heating of water and buildings can save both fuel and money for
utilities and their customers in such areas. This appears to be the case
in the specific instance of the Shetland Islands. The main obstacle to
direct involvement or investment by electrical utilities in solar heating
is likely to be institutional rather than technical.

REFERENCES

Bartholomew, D M L (1984). Passive solar house design in the UK.
    Proceedings North Sun '84 Conference, Scottish Solar Energy Group,
    Edinburgh.
Bossong, K. (1982). Citizen concerns with the solar energy/utility
    interface. Energy, 7, number 1, 141-153.
British Standards Institute (1980). BS 5918, London.
Ford, B H and Everett, R. (1985). Measured performance of attached
    conservatories. Proceedings of Conference Greenhouses and Conservatories
    UK-ISES, London.
MacGregor, A W K. (1979). Solar cladding: Insulation plus insolation.
    Proceedings International Solar Energy Society Congress, 1665-1668,
    Atlanta.
North of Scotland Hydro Electric Board, (1983 and 1985).
    Report and Accounts, Edinburgh.
Reynell, M J W (1984). Torbay hospital solar energy project and its
    predicted performance at high latitudes. Proceedings North Sun '84
    Conference, 205-222, Scottish Solar Energy Group, Edinburgh.

Solar Powered Services for Remote Aboriginal Communities

W. L. James

Solar Energy Research Institute of Western Australia,
PO Box R1283, Perth, Western Australia 6001

ABSTRACT

This project entails the design and construction of a mobile solar powered services system for small, nomadic aboriginal communities in Australia.   The services are incorporated in a 6 metre shipping container and include radio communication, water pumping, refrigeration, outdoor lighting and power for the operation of small power tools, TV and video recorder.

The first system was installed for the Ngurawaana Community in Western Australia and has been operating flawlessly since March 1985.

KEYWORDS

Photovoltaics; Australia; economics.

INTRODUCTION

Aboriginal communities in Western Australia, South Australia and the Northern Territory have been moving back into their traditional homelands since the 1970's. Their election to move away from townships and Missions has created a need for basic services.

Townships and Missions in Western Australia are presently powered by diesel generator sets supplied and maintained by the State owned utility.

The maintenance of equipment and the transport of fuel to some of these townships and missions is difficult and costly due to the large distances involved, the rough terrain and extreme weather conditions.

The dispersion of aboriginal groups into smaller communities, moving to even more remote areas has made the utilities, aboriginal groups and administrative authorities search for more practical solutions.

The State Energy Commission of Western Australia and the Department of Aboriginal Affairs, approached the Solar Energy Research Institute of Western Australia (SERIWA) to see if solar energy could prove a viable option. After several meetings with community representatives and advisers, it was concluded

I5.1

that it would be practical to design a solar powered system that could provide basic services rather than just electric power.   These service systems had to be relatively mobile as the communities are, by nature, nomadic.

A list of basic services for immediate consideration was established.

(a) Communications: The provision of reliable power to VHF transceivers for communicating with the Flying Doctor and other vital services.

(b) Water: The provision of pumping facilities to maintain existing potable water storage.

(c) Refrigeration: Provision of chilled and frozen storage for bulk meat, vegetables and dairy products.   Refrigeration could also be used for storage of medication, vaccines, etc.

(d) Lighting: Provision for area lighting.

(e) Battery Charging: For the charging of batteries for vehicles.

(f) TV/Video: Provision of power for a communal TV/Video set.

The prototype system was designed for the needs of the Ngurawaana Community located in the Pilbara region of Western Australia on the 21st parallel south, 110 km east of the town of Karratha.   This community had adequate water supplies in the form of a windmill, therefore the excess capacity was used in providing more refrigeration capacity.

SYSTEM DESCRIPTION

The prototype 'Solar-Pack' consists of a centralised facility based on an insulated cargo container with a roof mounted photovoltaic array, as shown in Figure 1.   The container forms a robust, transportable shelter for the associated electrical equipment.

The photovoltaic array is fixed to the roof of the container and consists of 750 Watts of Mobil Solar Corporation ribbon solar cells.   The container also holds a 19 kWh battery bank.   The batteries are Dunlop lead-acid, tubular traction type.   System voltage is 24 V DC.

An electronic controller prevents overcharging of the battery bank during periods of excess solar radiation while also ensuring that non-essential loads are disconnected during periods of poor solar radiation.

The prototype 'Solar Pack' was designed specifically for the Ngurawaana Community and provides the following services:-

(a) Communications

The system provides DC power to a 100 W VHF transceiver.   This facility replaces the existing low power transceiver already in use within the community.

(b) Refrigeration

The system supports four chest-type refrigeration units with a total capacity of 600 litres.   One of them is a commercial unit designed for photovoltaic applications.   This unit uses a single DANFOSS compressor.

The remaining three units were custom designed and built to SERIWA's specifications.   The 150 litre cabinet incorporates two identical DANFOSS compress-
or units operating in tandem with a single thermostat. This doubling of the
cooling capacity allows the unit to rapidly freeze a large joint of meat
despite high ambient temperatures and provides security of supply in case
of failure of one compressor - refrigeration system.

All the fridge cabinets are manufactured with high levels of insulation
(minimum 75 mm of polyurethane) and hence exhibit improved efficiency compared
to conventional domestic units.

The refrigeration equipment is configured as a refrigerator or freezer
by presetting the thermostat.   The prototype 'Solar Pack' was configured
with 50% fridge and 50% freezer capacity.

(c) Lighting

Fluorescent lighting was provided in the form of 20 W fittings which were
modified to accept fluorescent inverters.   This modification allows the
fitting to operate directly from a 24 VDC supply. Lighting was provided
in the container itself, in an adjoining office building and outside nearby
buildings.

(d) AC Loads

An inverter was provided to convert the low voltage DC power provided by
the system to mains voltage AC power (240 V - 50 Hz) for use with a conventional
television, video recorder and small power tools.   A mains powered battery
charger provides charging for 12 V automotive batteries.

The main power supply is provided in DC form to eliminate the reliability
and efficiency problems of inverters, as well as to contain the use of
large AC powered appliances which is common practice when a diesel generator
set is provided as a power supply.   The uncontrolled increase in load
generally ends up in an undersized diesel generator set or the selection
of a large set which continuously runs at low power level with the consequences
of poor fuel efficiency and cylinder "glazing".

DEPLOYMENT OF PROTOTYPE

The completed power supply system was transported to the Ngurawaana Community
campsite near Millstream, Western Australia on 8 March, 1985.

The main storage battery and refrigeration equipment are fixed permanently
to the walls and floor of the container and hence can be transported without
further modification.   The internally mounted transceiver and inverter
equipment are packaged separately and stored within the container for transport.

The container is normally mounted on two 4.5 m lengths of 203 x 76 mm steel
C-channels and bolted to the container sides to ensure stability in high
wind speeds.   During transport, these channels are stored within the
container and are fixed to the floor.

In its operating position, the photovoltaic array extends the vertical
height of the system approximately 500 mm above the container roof.   If
height is critical the array may be stowed parallel to the roof whereby
the extra height is reduced to 100 mm.   The overall height with the array
extended is 3 metres.

The system is lifted using four manually operated container jacks. A truck with a 6 m tray can be driven beneath the raised container with approximately 300 mm clearance between the tray and jacks on each side. The total weight of the system including container, jacks and support channels is 4800 kg.

Unloading is carried out in a similar fashion and takes a team of four approximately three hours to complete.

In transit, the 'Solar Pack' was subjected to very rough roads and rocky terrain during which the array was always in the extended operating position. On arrival, there was no apparent damage to any of the equipment. The refrigerators and freezers, although disconnected from the power supply, maintained the food frozen during the $2\frac{1}{2}$ days travel time.

External cabling was installed between the container and the store building and a nearby dwelling. This external cabling operates the lighting and will allow the transceiver to be installed in the store building when required.

Installation time from truck arrival to test and operation of the 'Solar Pack' was completed in one day. External underground wiring to a store and for outdoor lighting took two days due to the rocky terrain. Jack hammers had to be used for cable trenches.

The refrigeration equipment was configured as 300 litres of refrigerator and 300 litres of freezer space. The inverter was tested with hand tools and a television/VCR and operated reliably. The transceiver was tested in the morning of March 11th during the scheduled transmission by the Royal Flying Doctor Service in Port Hedland (4030 kHz). The transceiver showed considerably improved transmission and reception performance compared to the existing 25 Watt 12 V set.

Mr Greg Tucker of the Ngurawaana Community was briefed on the operation of the system and arrangements were made for the periodic recording of system performance via the panel meters on the controller enclosure.

SYSTEM PERFORMANCE

A member of the Ngurawaana Community arrived during the final week of construction to familiarise himself with the system and to assist with the final assembly. This proved to be a valuable experience as it has helped to establish a relationship with the parties concerned.

The 'Solar Pack' was tested at SERIWA's Solar Research Centre in Perth for a period of two months prior to delivery.

The design loads for the system are described in Table 1. In order to test the system for daily energy delivery, a constant current load of 10A was placed on the system for a continuous period of five days corresponding to a total daily load of approximately 6 kWh. The system operated during this period without load shedding despite maximum temperatures of 35-38°C. During this period, the battery voltage remained between 1.9 V/cell and 2.3 V/cell.

Instantaneous system performance can be monitored using the panel meters mounted on the front door of the controller enclosing. Normal array currents should range between 20-30 Amps under 1 kWm^{-2} insolation.

As can be seen from Table 1, the actual sustainable load is larger than the

maximum design load and hence load shedding should not occur except during
extended periods (4-5 days) of poor insolation.

TABLE 1

Load Estimates - Yearly Average

		Daily Load
(a) Refrigeration		2.5 to 3.5 kWh
Total Capacity	600 l	
Compressor Power	380 W	
(b) Lighting		1.0 kWh
Total Lighting	80 W	
12 hours/day		
(c) Communications		
Transmit Power	100 W	0.1 kWh
1 hour/day		
(d) AC loads		
Average power	50 W	0.4 kWh
5 hours/day		
	Total	4 to 5 kWh/day

This system has operated continuously since installation in early March.

A weekly contact is made with the Community and   SERIWA via a radio-
telephone patch with the Royal Flying Doctor.

Panel meter readings, battery specific gravity and general weather and
operating conditions are relayed. No abnormal readings have been noticed,
the lowest Specific Gravity reading being 1180 which is above the 75% charge
level of this type of battery.   Only one high value of 29 V at the battery
has been recorded at which time the controller switched the array off.   This
is a good indication that the array power is being well utilised and that
the system sizing is adequate so far.

Battery electrolyte has been topped up only once in the six month period.

The refrigerators and freezers are being well utilised with an average of 3
of the 4 units full.   The freezers were designed to cope with fast freezing
of freshly killed animals. A problem has occurred with the dripping of blood
which is a nuisance if plastic bags are not used.   However a regular cleaning
of freezers is effected when running low on food.

Vegetables are being stored and consumed more regularly now. However,
sometimes these have been put in the freezers rather than in the refrigerators
and have been occasionally damaged by frost.

The general comment by the community is that they are happy with the system.
Indications are that more area lighting would be desirable for the more
distant dwellings.

SYSTEM ECONOMICS

(All values in A$, 1A$ = 0.7 US$)

The costs of major system components are outlined in Table 2.   A labour
component of $3000 has been included to represent the additional costs
involved in commercial production.   It should be noted that some components'
costs, especially that for refrigeration, could be reduced if the equipment
were produced in larger quantities.   However, profit margins necessary for
a commercial production will increase the costs.

TABLE 2

Photovoltaic System Costs

Photovoltaic array (750 Wpk)	£7,500
Battery (800 Ah, 24 V)	2,300
Electrical fittings	1,500
Refrigeration	5,660
Inverter	760
Transceiver and antenna	2,300
Container (second hand)	1,600
Miscellaneous hardware	400
Labour (150 hours at $20 per hour)	3,000
TOTAL	$ 25,020

Only the power supply plant is considered in the economics as the rest of the
system should remain the same.

Assuming that the battery will require replacement after five years and the
balance of the system after 20 years (with no residual capital value) then
the annual running costs of the system would be approximately $0.97/kWh.
This assumes an average energy delivery of 5 kW h/day and an interest rate
on borrowed capital of 12%.

The existing variable load (180 W average) is not compatible with conventional
diesel driven generation equipment.   The smallest practical generator size,
which would be around 2 kW, would only be loaded to 10% of rated load for
much of the time.   A generator-set operating under these conditions would
exhibit extremely poor fuel efficiency (less than 1 kWh/litre for diesel)
and correspondingly poor reliability.

A better alternative is the use of a petrol driven battery charger in conjunction with the existing battery bank.   This would result in a cost of
approximately $1.33/kWh.   Clearly, the use of photovoltaics is the most
economic solution.

The use of photovoltaics produces an immediate economic advantage compared
to the petrol based system.   This economic advantage will become more significant as the real cost of petrol increases in the future.

CONCLUSIONS

A photovoltaic power supply system has been constructed and deployed at the
Ngurawaana Community campsite near Millstream, Western Australia.

The system provides reliable DC power to communication, refrigeration and lighting equipment and limited AC power for handtool, television and video equipment.

The system is easily transported and can withstand rugged road conditions without apparent damage to any of the equipment.

The solid container can withstand cyclonic weather conditions and could prove a temporary shelter when these occur.

The system requires minimal maintenance and should significantly enhance the quality of life for remote communities.

A second unit is being constructed for a community located in the Great Sandy Desert.

SERIWA is organising and directly supervising the construction of these and future units.

TOPIC J
Building Design

Demonstration of Energy Efficient Housing — the Relevance of the BRECSU Programme to Rural Communities

J. R. Britten

Building Research Energy Conservation Support Unit,
Building Research Establishment, Garston, Watford, Herts,
UK

1. INTRODUCTION

The Building Research Energy Conservation Support Unit (BRECSU) is one of the two support units which manage a programme of research, development and demonstration on behalf of the Department of Energy's Efficiency Office (EEO). (The other unit is the Energy Technology Support Unit (ETSU) which is perhaps better known to this audience for its work on alternative energy sources, including the utilisation of solar energy in buildings).

BRECSU shares with ETSU the responsibility for operating the EEO's Energy Efficiency Demonstration Scheme. This scheme is intended to hasten the take up of energy efficient technologies throughout all sectors of the economy, and BRECSU was formed in 1981 with the expressed purpose of extending the demonstration scheme into the housing sector.

BRECSU is an integral part of the Department of Environment's Building Research Establishment, and so not only is associated with BRE's teams of research workers who are currently studying many aspects of energy efficiency in housing, but furthermore is heir to the traditions and experience of BRE, which stretch back over 60 years. This assists BRECSU staff to consider the wider aspects of construction, maintenance, and occupancy when making assessments of novel energy saving proposals.

2. THE DEMONSTRATION PROGRAMME

The broad objectives of the demonstration programme, and the criteria for support of projects, are set down in a number of publications[1][2] from the EEO and will not be described at length in this paper.

It is sufficient to note that any building owner who wishes to install a novel, yet cost-effective, energy efficiency measure in his property, but who is concerned about the uncertainties of its performance or the potential risks involved, may apply for a grant to cover 25% of the cost of the installation. In return for this grant, he must provide access to the property so that an independant consultant can monitor and assess the full effects of the measure. The results of this assessment are then published widely in order that owners of similar buildings are aware of the true benefits of energy efficiency.

J1.1

When applied in the housing sector, the definition of 'novelty' is extended
to cover not only devices which are technologically novel, but also design
principles and construction methods which are not yet widely applied in
house-building. Projects may relate to improvements in the energy
efficiency of the building envelope, of the systems for providing space and
water heating, or of other energy consuming appliances such as cookers. In
an earlier paper (3) BRECSU identified almost 100 potential demonstration
projects in the domestic sector alone, and the remainder of this paper will
discuss the implications of these projects upon housing for rural
communities. (All those projects involve groups of properties, as it is not
usually practicable to accept single dwellings for monitoring.)

2.1 NEW HOUSING

Some of BRECSU's most successful projects have been in the area of new
house construction - although as will be explained later this area is of
less importance nationally than that of improvements to the existing stock.
Two schemes in particular have shown how an understanding of the values of
traditional house design, taken with an acceptance of modern materials and
design aids, can lead to the construction of houses which cost little more
than their neighbours yet use only half as much energy for space heating.

The traditional remote cottage, Welsh farmhouse or Scottish croft, was
built of local heavyweight materials, had thick walls, small windows, and
possibly only one door on the sheltered side of the building. It may also
have had a very well insulated roof of local thatch. As a result there
were only small heat losses through the structure, and the small building
could be kept at a comfortable temperature by a single open fire, having a
central flue which transmitted warmth to the other rooms.

More recently these traditional principles played a lesser part in design,
as newer construction materials enabled load bearing walls to be of slimmer
section, windows became larger for reasons of visual amenity, and the
(relatively) low fuel prices meant that designers paid little attention to
the energy efficiency of their schemes. Blocks of flats, built of load
bearing uninsulated concrete, heated by electric underfloor systems, and
with large windows and draughty access balconies represent a totally
different approach to housing than that exemplified by the small croft.

But a BRECSU project in Manchester has shown how the traditional principles
can be reapplied to current local authority family housing. Insulation is
applied all around the dwelling - in wall cavities, beneath the floor, and
in the loft - at thicknesses greater than required by current regulations.
Windows are well fitting, (but with trickle ventilators to reduce the risks
of condensation), are dual glazed, and are smaller on the North side of the
house than on the South. Projecting eaves reduce the risk of overheating
in summertime. Doors are draughtproofed, and a draught lobby is included
within the internal plan of the house (rather than as a projecting porch
which would lead to increased heat loss as well as costing more to build).
Although the heating is no lon ger by a single open fire,yet the central flue
is retained as an integral part of the heating system design.

None of these measures is new, and anyone could have applied them, but by
careful attention to detail at both the design and construction stages
these houses were built at a cost only £300 more than their standard
designed neighbours, but they now have fuel bills of £97 a year less (4).

Similar principles were employed on a scheme in South London, where the designers have embodied their approach to integrated housing design in four simple rules(5). The first of these rules points out that a given amount of insulation can have more effect when distributed throughout the structure (eg in walls, floors and roof) than if merely positioned in the few easy-to-get-at places such as the loft. Orientation and window sizing are also considered in the 'rules', and the designers give great attention to draughtproofing. In addition to supplying draughtproofing at the obvious openings of doors and windows, the specification includes the caulking of all service entries, pipe runs through the ceiling and light fittings, in order to reduce still further the adventitious infiltration rate.

These measures have resulted in a group of houses which have an attractive appearance, and have space heating energy bills of less than £100 per year. The design also incorporates a solar water heating system (using the 3-tap thermosyphon principle) and although this is a relatively inexpensive installation, which does provide substantial quantities of solar heated water in favourable conditions, yet it is notable that its cost-effectiveness is much less than that of the more straightforward fabric measures.

These two schemes, in different parts of the country, show that heat losses through the building fabric can be substantially reduced, without incurring great expense, if both the designer and the builder adopt an integrated approach to design and construction. This is something more than just 'good practice'. Traditional ideas, applied here in an urban context, can be handed back with improvements to the remote areas from which they originated.

2.2 REHABILITATION SCHEMES

However new housing can only have a limited effect upon the total national energy demand. Within the United Kingdom there are at present some 20 million households, and new houses are currently being built at the rate of approximately 200,000 per year. When coupled with the trend towards a greater number of smaller households, and with a housing demolition programme of about 50,000 units per year, it can be seen that our houses probably need to last for at least 100 years; the existing stock is therefore the real key to energy efficiency in the housing sector. For this reason BRECSU's programme gives a considerable emphasis to the energy efficient rehabilitation of existing dwellings.

Many millions of dwellings in country districts were built with walls of solid masonry construction, but without the thickness and mass of the traditional cottage construction. These solid walls have poor thermal insulation qualities, and are more difficult to insulate by retrofit measures than are (for example) cavity brick walls. Many of BRECSU's projects are concerned with means of improving the insulation of solid walled houses. Some of the schemes are in Victorian properties, in urban areas, but one of the earliest projects took place in rural Gloucestershire. Although superficially the results were discouraging, they are worth describing because of their implications for the assessment of similar schemes.

The houses were of a concrete post and beam construction, common in the West Country and similar in principle to prefabricated systems which have been used for rural re-housing schemes in several parts of the world. Calculations showed that the walls had a poor U-value (around 2.4 W/m^2°C) but that their performance could be improved by some 50% by the application of a proprietory system of external insulation and rendering. This improvement would result in a substantial energy saving, and give an adequate return on the investment. In fact a carefully controlled monitoring exercise(6) involving 60 houses in three different villages, showed only a small absolute energy saving, which was certainly not worthwhile in terms of the cost of the insulation. But this initial finding took no account of the increased comfort obtained by the occupants. Their houses became noticeably warmer, they suffered less from condensation, and so in turn their redecoration and maintenance costs were expected to be reduced. The external appearance of the houses was improved, and as a result of all these factors the residents began once more to be satisfied with their houses. An estate which was in danger of becoming 'hard-to-let' became by contrast a popular place to live.

It is not uncommon for the potential savings from energy efficiency to be taken as an increase in comfort, and this example shows that criteria of financial cost effectiveness have to be interpreted carefully when dealing with housing improvements, and BRECSU's monitoring programmes always include an assessment of the sociological factors consequent upon any energy measure. Such factors could include the possible consequences of a breakdown of rural communities, if obsolescent housing is left to decay, although it has to be realised that the demonstration scheme is an 'energy' demonstration scheme and not a 'social housing' demonstration scheme.

2.3 HEATING SYSTEMS

The examples quoted so far have been largely concerned with the building envelope, but BRECSU's programme also deals with the efficiency of heating systems. One project which is now in progress in an inner city area could eventually provide useful information for remote settlements, since it involves the use of small scale combined heat and power systems (CHP) to service part of a local authority estate. These mini CHP schemes have not yet been widely used for housing, although there is an increasing body of information available about their performance in commercial buildings. Family housing has a different demand pattern from commerce (both for heat and for power) and the BRECSU project will show how well this demand profile can be met by the CHP systems.

It is expected that even in this urban situation, the CHP scheme will provide heating services more economically than the previously (conventionally boilered) system. In a rural community, not connected to grid electricity supplies, the system could be even more effective.

CHP systems are more suited to groups of properties than to individual dwellings, but another of BRECSU's projects is monitoring the performance of an improved type of domestic central heating boiler, which has an efficiency considerably greater than that of conventional appliances. Originating in Germany and the Netherlands, these 'condensing' gas boilers have recently been introduced to the British market, and one product is already approved for use with liquid propane gas for districts outside the

gas supply area.   The boilers extract the latent heat from the flue gases,
cooling them below the dew point, and as a result they have bench
efficiencies which are up to 10 percentage points greater than the best
current models;   equally importantly, their part load efficiency is also
much improved.

The BRECSU programme is studying the full effects of their introduction,
including householder's reactions to condensate drains, exhaust plumes, and
lower radiator surface temperatures, as well as measurements of fuel saving
and actual efficiencies.   First results are encouraging, and wider trials
are about to take place, including some projects in rural areas.

The introduction of these boilers could have a marked effect upon domestic
energy consumption - both individually and nationally - within the next few
years, as existing boilers are replaced.   The condensing boiler is one of
the few energy efficiency devices that is capable of being installed in
both new and existing properties, in rural or urban areas, and in the
private and public sectors. Its impact upon rural communities is likely to
be as great as in the cities, and its cost effectiveness may even be
increased in remote districts, where fuel is more expensive.

3. CONCLUDING REMARKS

The examples presented earlier have given a brief outline of BRECSU's
programme of demonstration projects in the domestic sector.   Although most
projects are sited in urban areas, and none have yet taken place in truly
remote regions, yet nonetheless there are benefits to be obtained by
building owners and occupiers in all parts of the nation. When seeking new
projects, or when assessing proposals for support, BRECSU gives attention
to regional and geographical factors, and the overall strategy recognises
the importance of a geographical spread of projects.   Measures which are
cost effective in one part of Britain may be uneconomic or inappropriate in
another region; and schemes which are acceptable in a city may not accord
with traditions or standards in a country district.   It is therefore
helpful to have a range of projects in a variety of situations.   BRECSU
also seeks to acquire projects in different parts of the country in order
that potential replicators can more easily visit demonstration sites, and
see for themselves the benefits that can accrue from the application of
newer technologies or designs. Display material at the different locations
can indicate the particular advantages that are appropriate to different
audiences and different climates.

As yet, few proposals have been received from remote settlements, but
schemes submitted on behalf of rural communities in any part of Great
Britain will be given sympathetic consideration by BRECSU when being
assessed against the criteria for support.

Communities which do not yet have projects eligible for support can still
benefit from the demonstration scheme, by using the results from projects
that have been completed elsewhere.   The BRECSU enquiries bureau can
distribute copies of short profiles (giving brief descriptions of current
projects) and of full reports (giving monitored results and conclusions
from completed schemes).   Additionally BRECSU staff can give more detailed
presentations to groups of designers or building owners faced with the task
of making energy efficient improvements to housing in rural communities.

4. ACKNOWLEDGEMENTS

The work described forms part of the research programme of the Building
Research Establishment, and this paper is published by permission of the
Director, BRE.

BRECSU manages its programme on behalf of the Department of Energy, but the
views expressed in this paper are not necessarily those of the Department.

REFERENCES
1.   'Financial assistance for demonstrating the efficient use of energy.'
     Energy Efficiency Office, London, 1984

2.   'What you need to know.' Energy Efficiency Office, London, 1984

3.   'Energy   Efficiency   Demonstration   Scheme   -   A   Review.'
     HMSO, London, 1984

4.   'Low energy houses in the City of Manchester.' Report F37/84/89 by
     UMIST, BRECSU 1984

5.   'Energy Efficient Housing - a demonstration of the integrated
     approach.' Energy Efficiency Office and South London Consortium,
     London, 1985

6.   'External wall insulation applied to Woolaway system houses.' Report
     F/27/83/88 by PA Consultants, BRECSU 1983

All references except (3) are available from the BRECSU Enquiry Bureau,
free of charge.

Autonomous Electrical Supply Systems Based on Renewable Energy Sources — Results from a Simulation Model and Experiences with a Small Scale System

H. G. Beyer, H. Gabler, G. J. Gerdes, J. Luther and W. Schmidt

Fachbereich Physik, Universität Oldenburg, Postfach 2503,
D–2900 Oldenburg, FRG

ABSTRACT

By combining today's techniques of energy conversion and storage it is possible in principle to construct energy supply systems based to a high degree
on renewable energy sources. At Oldenburg University a small laboratory
building is in operation whose energy supply is covered by solar radiation,
wind energy and to a small extent by conventional fuel. We report on simulation calculations for such systems using experimentally validated models for
converters and storage devices. In particular the influence of system parameters such as size of different converters and battery capacity on the renewable fraction and the energy pay back time of the whole system is discussed.

KEYWORDS

Renewable energy; wind energy; solar energy; renewable energy systems; stand
alone systems; energy pay back time; system optimization.

INTRODUCTION

Renewable energy systems have to be designed for the meteorological conditions at a site as well as the time pattern and amount of expected consumption. For today's applications the figure of merit for any optimization of
these systems will be investment and operating costs; in order to assess
large scale application in the future it is reasonable to look at energy pay
back times.

We have performed optimization calculations with respect to energy pay back
time for an electrical supply system characterized by a non-flexible consumption pattern and the utilization of three energy sources: wind energy,
solar radiation and conventional fuel. It is of typical structure for stand
alone applications of renewable energy sources.

317   J2.1

The system characterized by Fig. 1 supplies electrical energy to a small
university building (the 'Energielabor') used for teaching and laboratory
work by the departments of natural sciences. The 'Energielabor' is not connected to the public utility supply.

On the basis of energy flow balances measured in 1984 the characteristics of
the energy converters and storage were modelled. With the input of meteorological data taken on site over previous years subsequent simulation runs
using time steps of one hour yield system performances over yearly periods.

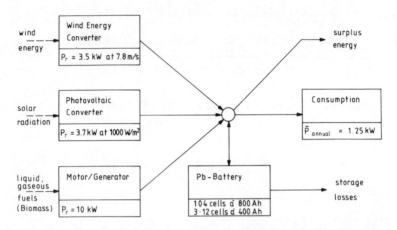

Fig. 1. Typical energy flow chart for a stand alone elec-
         trical supply system. System layout for a given
         site and given energy consumer may be defined by
         three parameters: storage capacity, size of wind
         and size of solar energy converters. Surplus
         energy which may be generated at times of high
         renewable energy availability is of no use to
         the system owing to limited storage capacity.
         The data given in the Fig. are the rated values
         for the system operated at Oldenburg University.

COMBINED SYSTEMS

The performance of renewable energy systems may be characterized by the
fraction (F) of produced energy that is derived from renewable sources (F =
1.0 if no motor fuel is consumed); surplus energy must not be taken into account when calculating F.

System configuration is best specified by the gross renewable production
(PRG) (including surplus) in units of consumed energy and the ratio of wind
to solar produced energy (W/S). These dimensionless values are generated by
linking the converter characteristics with the meteorological data at the
site of interest.

We have investigated (Beyer and others 1984, Gabler 1985) renewable frac-
tions (F) as function of production PRG, storage capacity BAT and ratio W/S
(Fig. 2). As expected the renewable fraction rises with growing production
PRG or capacity BAT. Combining wind and solar energy yields better system

performance for all levels of renewable production or storage size than wind or solar conversion alone. Other investigators (Blegaa and Christiansen 1981, Blok 1984) reported the same effect for large electrical power networks incorporating wind generators and solar cells.

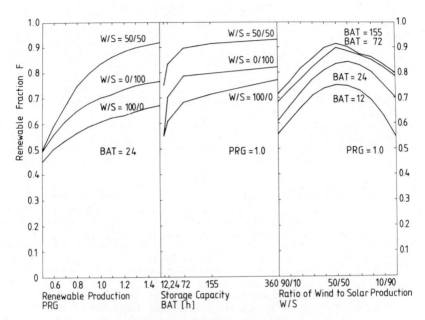

Fig. 2.   Renewable fraction of energy production as function of renewable production, storage capacity and ratio of wind to solar produced energy. Systems combining wind and solar power show better performance than systems using wind or solar exclusively. All calculations were made for the meteorological conditions at Oldenburg and the time pattern of consumption in the Energielabor. The 'irregularities' in the graphs are due to the control strategy of the motor/generator system used in the simulation: when run the motor/generator charges the battery always to half its rated capacity.

ENERGY PAY BACK TIME

The renewable fraction is in itself no criterion for system optimization since it does not measure the extent of equipment involved in order to achieve a given system performance.

Hence we define an energy pay back time as optimization criterion: it is given by investment in terms of final energy consumed in establishing and operating the system divided by the useful mean power output (i.e. generated power minus storage losses and surplus energy). The nonrenewable fuel consumption (being final energy itself) is integrated over the system's life-

time and treated as energy investment at the beginning of system operation.
Thus every kWh of useful energy generated after the end of the energy pay
back period substitutes exhaustible resources. Fig. 3 gives an example for
system optimization with respect to energy pay back time.

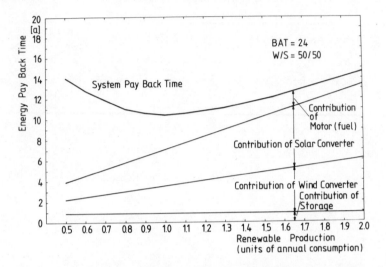

Fig. 3.   Energy pay back time is a function of annual renewable energy production. Storage size and ratio of wind to solar produced energy are kept
constant in this calculation. Variation of the
renewable energy production leads to a minimum
in energy pay back time: with growing contributions from wind and solar converters the contribution of motor fuel decreases. Production of
(un-utilizable) surplus energy becomes important
if the renewable energy production exceeds consumption appreciably. The absolute values given
at the ordinate correspond to conservative assumptions; moderate changes in technology could
lead to considerably lower pay back times in
the future.

It is interesting to note that in optimizing the pay back time the storage
capacity plays a crucial role. Not only has the energy investment for the
storage to be repayed, but storage capacity also determines the fraction of
generated energy put to final use within the system. This should outline the
relatively complex optimization paths involved and shows that energy pay
back times may only meaningfully be defined for whole systems.

OPTIMIZED SYSTEM CONFIGURATION

Calculations show that for the meteorological conditions at Oldenburg (see
below) about equal shares of solar and wind produced energy result in better
system performance not only with respect to the renewable fraction but also

with respect to energy pay back time. The optimized configuration for a site with 50 % higher annual wind speeds (6.0 m/s) would be (using today's technology) a pure wind system without use of solar energy.

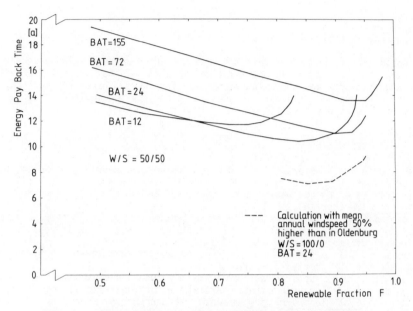

Fig. 4.   Energy pay back time is plotted versus renew-
          able fraction: for a fixed ratio W/S the mini-
          mal pay back time is found for capacities near
          24 h. Systems should be configurated with high
          renewable fractions even for the moderate mete-
          orological conditions at Oldenburg. For sites
          with 50 % higher wind speeds, energy pay back
          time is considerably lower.

Given the ratio W/S (50/50) and a fixed storage capacity, energy pay back time can be studied as a function of the renewable fraction. Fig. 4 shows that a minimum of energy pay back time is achieved at high renewable fractions for all reasonable storage capacities. An absolute minimum of energy pay back time for the systems under consideration is found for storage capacities of about 24 hours.

The meteorological conditions at the site of our experimental building and the data of an optimized system are the following:

meteorological data:
mean annual wind speed   4.1 m/s
mean annual insolation   110 W/m²

system layout:
storage capacity (hours of mean consumption)   24 h
ratio of wind to solar produced energy (annual average)   50:50
annual ren. energy prod. (in units of ann. consumption)   1.0

system performance:
renewable fraction   0.84
energy pay back time (conservative estimate)   10.4 years

We would like to point out that the data presented above should be understood as an example for an optimization calculation. They are valid only for
our specific site and the assumptions made for energy investment in today's
standard technology. Moderate changes in technology should lead to considerably lower pay back times in the future.

REFERENCES

Beyer, H. G., H. Gabler, G. J. Gerdes, J. Luther, E. Naumann, M. Nolte, W.
    Schmidt and R. Steinberger (1984). Modelling autonomous energy supply
    systems based on renewable energy sources - an example -. Proceedings of
    the 5th ISF (Berlin), 1074-1078.
Blegaa, S. and G. Christiansen (1981). Potential for a Danish power system
    using wind energy generators, solar cells and storage. The International
    Journal of Ambient Energy, 2, 223-232.
Blok, K. (1984). A renewable energy system for the Netherlands. Proceedings
    of the 5th ISF (Berlin), 90-95.
Gabler, H. (1985). Optimierung der Auslegung eines regenerativen Energie-
    systems - Modellrechnungen zum Energielabor in Oldenburg. Ph. D. Thesis.
    University of Oldenburg. (in German)

Energy Analysis for a Medium-sized Creamery in Scotland: a Strategic View of the Potential for Energy-cost Savings by Application of Conventional and Alternative Technologies

M. J. Harwood and J. K. Jacques

Technological Economics Research Unit, University of
Stirling, Stirling FK9 4LA, UK

ABSTRACT

The paper describes how basic exergy analysis might be used to evaluate the potential contributions of fossil and renewable (e.g. solar) energy resources towards a more energy efficient operation in a creamery.

An index of energy quality is proposed which could be used as a proxy for detailed exergy calculations. The energy quality index illustrates the current mismatch between energy quality of supply and energy quality of demand.

The usefulness of simplified exergy analysis is studied and the conflict of interests between energy cost minimization and exergy loss minimization is described.

KEYWORDS

Exergy analysis; energy quality index; renewable energy sources; energy cost minimization; energy management strategy.

INTRODUCTION

An energy survey was carried out at a medium-sized creamery in Scotland whose products at the time of study included special cooking cheeses, cottage cheese and cream. The company had recently moved their creamery operation to another site and had taken over the existing services which were originally intended for milk bottling.

The aims of the survey were:

(a)   to study ways of improving the energy efficiency of the creamery.
(b)   to study the usefulness of simplified exergy analysis to the management of the creamery – and for other applications.

J3.1

(c)   to construct a practical index of energy quality which could be
           used as a proxy for exergy analysis.
(d)   to study how the company assessed its energy-saving investments
           and what implications this may have for (b).

ENERGY ANALYSIS AND EXERGETIC EFFICIENCY

A detailed on-site study of the process energy requirements and of the
current energy supply systems was carried out. Figure 1 shows the major
energy (and exergy) flows and losses for the steam and hot water supply
system.

The first law efficiency (η_1) relates the useful energy output to the total
energy input and was calculated as 55% for the steam and hot water supply
system. The major loss of energy is due to low boiler efficiency (caused
by excessive overcapacity on boiler plant).

However, as various authors have previously shown (Ahern, 1980; Gaggioli,
1983) the first law efficiency is misleading because the quality of the
energy lost is not considered. When analysing energy use it is important
to assess an energy flow's capacity to do work, as well as its heat content.

Fundamentally, exergy as defined by the Gibbs Helmholtz expression $\Delta G = \Delta H$
$- T.\Delta S$ reflects the availability of work from a given heat-change or heat
transfer process performed under carefully prescribed steady-state conditions,
in relation to a bench-mark standard state for the working substance or fuel.

The practical difficulties for the energy engineer in using the generalized
concept are manifold:

(a)   standard states bear no resemblance to real working conditions in
           specific situations.
(b)   'ideal' working engines (cf. Carnot cycle) are generally remote
           from practical high-rate machines; or from the needs of rapid heat
           transfer processing.
(c)   while efficient use of exergy is an accepted desideratum, the
           techno-economist is also faced with a need to optimize energy use
           and energy costs in terms of the opportunity values of his
           resources. From an energy conservation management viewpoint,
           energy cost minimization rather than minimization of exergy is of
           greater concern. This conflict of interests is reflected in the
           creamery example.

An averaged second law efficiency (η_2), which relates the minimum exergy
required to the actual exergy used, of 10% was calculated for the steam and
hot water supply system. This exergy efficiency figure is low because a
high quality energy source is providing for low quality energy demands.
This analysis agrees with earlier research (Twidell, Pinney, 1985), that for
a high exergetic efficiency the quality of supply should closely match the
quality of demand. In practice however, this will depend on local environmental factors and energy sources (wind, solar, hydro) and their adequacy
over the duty cycle.

An index of energy quality may be more useful for energy quality matching
and for illustrative purposes than detailed exergy calculations; although
this has been studied (Jacques, 1979; Twidell, Pinney, 1985), the energy
quality index has still to gain general acceptance.

                                                                    J3.2

The energy quality index now proposed differs from that of recent research by using the expression q = W/Q where q = quality, W = exergy or available work and Q = total heat input, for both the quality of supply and the quality of demand, giving more accurate figures for the quality of end use demands. Figure 1 gives the consequential index values, illustrates the reductions in energy quality within the plant and the low exergetic efficiency of the chill supply systems.

APPLICATION OF RENEWABLES TO INCREASE EXERGETIC EFFICIENCY

While solar energy can be used for high quality demands (Twidell, Pinney, 1985), solar may additionally be used for lower quality demands. The use of flat plate collectors to supply hot water would give an exergy efficiency of approximately 75%. Wind power (and solar electricity) would increase the exergetic efficiency of the electricity supply system as a whole. The creamery chill supply system has a low exergetic efficiency (see Fig. 1). Solar powered absorption refrigeration systems would increase η_2 by a factor of more than three and may well be technically and economically feasible in Southern European applications. A CHP installation would of course greatly enhance the exergetic efficiency, but would require a careful balancing between the cyclical demands for electricity and heat. Even with the relatively favourable 2:1 heat/electricity ratio overall, the economic return is poor (Payback Period greater than 20 years).

THE MANAGEMENT'S ENERGY-SAVING INVESTMENT STRATEGY

The firm's monthly energy cost totals approximately £11,000, 60% of which is the cost of fuel oil. Management were keen to reduce their energy costs and agreed to study a broad range of energy-saving investments. Table 1 gives the projects considered, the economic feasibility of each and the stage of development.

Projects were assessed by management on the basis of the limited Payback Period (PBP) criterion for economic feasibility. All projects agreed to date have a PBP of less than two years. However, PBP is not the best choice of economic criterion as it gives too much priority to short-term investments and leads to missed opportunities. The Nett Present Value (NPV) gives a more useful estimate of the worth of a project, since lifecycle costing of any investment (especially using renewable energy sources) is vital for reliable economic performance assessment.

Management were unwilling to use any criterion other than PBP and would not accept projects with a PBP of more than two-three years because of the particularly volatile fast-food market which the plant served. They felt that the position of the company was not sufficiently secure to justify more long-term projects. High interest rates were also a limiting factor on corporate investment strategy.

SUMMARY AND CONCLUSIONS

The true measure of energy efficiency (η_2) contrasts sharply with conventional energy analysis (η_1). Pursuing the consequences of exergy analysis calls for high capital expenditure and long-term investments. Current corporate energy-saving investment policy on the contrary only considers short-term projects with quick returns.

Exergy analysis should find application at the design stages of all industrial plant and should serve as a useful tool in national energy planning,
highlighting the true sources of inefficiency in industrial energy use.

The concept of indexing energy quality is discussed, with a view to the
development of a quantifiable management energy accounting model with
practical estimating capability.

TABLE 1   Energy-Saving Investment Possibilities

Project	PBP (years)	Project Development
Fuel Conversion (oil to gas)	1	√
Refrigeration Compressor Replacement (problem of overcapacity)	1-2	√
Heat Recovery - condensate	1	√
- hot air from cooling towers	>5	x
Pipe Lagging	<1	√
Heat Pumps - air conditioning	n.a.	?
Phase Lag Capacitor	1-2	√
Hand Controls for Hot Water Hoses	<1	√
CHP	>20	?
Wind Power - to supply electricity	>6	x
Biogas - methane from whey	n.a.	?
Solar Water Heating	>20	x

Key: √   Agreed

          ?   Under consideration

          x   Abandoned

          n.a.   Data not yet available

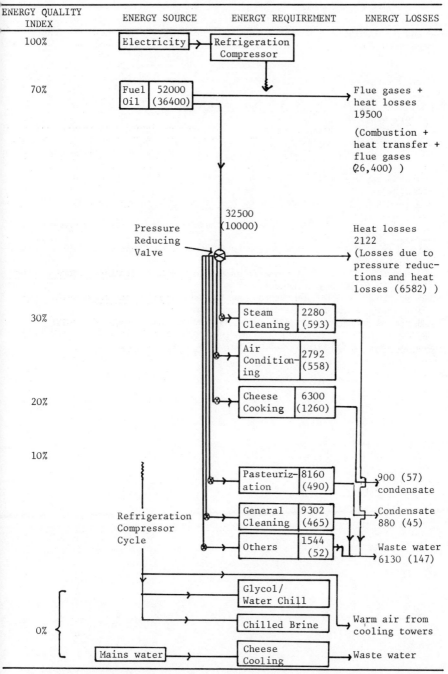

Fig. 1 Major energy and (exergy) flows and energy quality index
(Energy values in MJ/day; exergy values in (MJ/day) bracketted)

ACKNOWLEDGEMENTS

The authors would like to thank the management of the creamery for their
co-operation and the S.E.R.C. for their financial assistance.

REFERENCES

Ahern, J.E. (1980). The Exergy Method of Energy Systems Analysis. J. Wiley
& Sons, New York.
Borel, L. (1976). Energy Economics and Exergy Comparison of Different
Heating Systems Based on the Theory of Exergy. In E. Camatini, T. Kester
(Eds.), Heat Pumps and their Contribution to Energy Conservation.
Noordhoff - Leyden. pp., 51-96.
Brinkworth, B.J. (1972). Solar Energy for Man. The Compton Press,
Salisbury.
Dixon, J.R. (1975). Thermodynamics I: An Introduction to Energy. Prentice
Hall Inc., New Jersey.
Duffie, J.A., and W.A. Beckman (1974). Solar Energy Thermal Processes. J.
Wiley & Sons, New York.
Gaggioli, R.A. (1983). Efficiency and Costing. ACS Symposium Series 235,
Washington D.C.
Jacques, J.K. (1979). Watt Committee on Energy Consultative Council Report
No. 6. In Mech. Eng., London.
McVeigh, J.C. (1983). Sun Power. An Introduction to the Applications of
Solar Energy, 2nd ed. Pergamon Press, Oxford.
Twidell,J.W. and Pinney,A.A(1985). The Quality and Exergy of Energy
Systems, Using Conventional and Renewable Sources. In L.F. Jesch (Ed.).
Sun at Work in Britain, (March 1985). UK-ISES, London.

Energy Self-Sufficient House

S. C. Bajpai and A. T. Sulaiman

Sokoto Energy Research Centre, University of Sokoto,
Sokoto, Nigeria

ABSTRACT

This paper gives some technical features of a proposed energy self-sufficient house. All the energy requirements of the house are to be met using renewable energy resources within the periphery of the house. Traditional and passive architecture are to be used with some renewable energy systems integrated with the house design.

KEYWORDS

Energy, House, Architecture, Solar, Biogas, Woodstoves, Photovoltaics, Heating, Cooling, Cooking.

INTRODUCTION

Availability and accessibility to commercial energy in many rural and remote areas is limited. Inhabitants in these areas still rely on the traditional and local methods to meet most of their energy requirements. With the result, there has not been any significant improvement in the life-style of these people. Harnessing of locally available renewable energy resources for domestic, agricultural and industrial purposes using appropriate technologies can appreciably improve the conditions in these areas.

Some renewable energy systems for various domestic energy requirements have been locally fabricated and tested. Among them are: Solar cookers (Ahmed, Muazu and Sulaiman, 1984); Solar water heaters (Akujobi, 1984); Biogas plant (Fernando and Dangoggo, 1984); Efficient woodstoves (Ahmed and co-workers, 1985); and Passive air coolers (Bugaje, Saraki and Sulaiman, 1984). Some other imported systems have also proved their reasonable worth in the local climatic conditions (Bajpai, Rostocki and Sulaiman, 1985; Atiku, Bajpai and Sulaiman, 1985). Integration of all these systems with some simple energy responsive building design has also been tried at proto-type level (Bajpai, 1985). Based on all these experiences, a concept of energy self-sufficient house - a house which is to relay only on renewable energy resources within its periphery for all its requirements - has been conceived. It is planned to construct this house at the main campus of the University of Sokoto. This paper gives some technical features of the proposed house.

J4.1

HOUSE DESIGN AND CONSTRUCTION MATERIALS

The house has been designed for a floor area of 400 m^2. Fig. 1 gives the floor plan of the house and the surrounding facilities.

Sokoto, having a hot climatic feature most of the year, requires considerable amounts of energy for space cooling in the houses. Traditional architecture and some simple passive concepts have been combined to be used in this house to reduce its cooling load. Orientation and shape of living space in the house is designed to provide substantial sun control and cross-ventilation. Trees like neem and eucalyptus are to be planted near the south side wall of the living and bedroom zones, to provide shade.

Fig.2 gives the roof plan of the house. The roof on utility zone will be of three types - A lower level R.C. slab roof on conventional kitchen, a glass-plastic transparent roof on solar kitchen, and the remaining area having G.I. roofing sheets. All these roofs are to be slanted 15° towards south. An attic store facility is to be provided on the top of dining room, store and pantry. Some of the living space would be provided with clerestories for effective utilization of natural daylight.

Heat rejecting type of materials are to be used in the construction of the house, particularly in the living and bedroom zones. Such materials include cavity bricks, plywood and asbestos sheets. The house is to be painted with glossy white paint.

RENEWABLE ENERGY SYSTEMS

Major energy requirements of the house would be in cooking, water heating, lighting, domestic appliances operated with electricity, water supply, and cooling and heating of living space. All these requirements are planned to be met using various renewable energy systems.

Cooking

The house is to use three types of cooking systems:
A four-booster moderate size solar oven and a hot box to be placed in the solar kitchen. These systems are meant for low temperature boiling type of cooking requirements.

Two methane burners inside the conventional kitchen to be fed by a biogas plant of 8 m^3 gas per day capacity. This biogas plant is to be a floating gas-holder and continuous feeding type. Animal dung and poultry droppings obtained in the house are to be the raw materials for this plant. Methane burners are to be used for cooking in the evening and also for high temperature cooking during the daytime. The digested slurry of the biogas plant is to be used as fertilizer for gardens, animal feed crops and fuelwood plantations. Two efficient woodstoves (2-hole and 4-hole) are to be provided in the conventional kitchen with their chimneys upto roof level. These woodstoves will serve as standby to the above two cooking systems, particularly for harmattan and rainy seasons. Fuelwood for these stoves is to be obtained from the fuelwood plantations.

Water heating

A solar collector of 3.2 m^2 area mounted on the R.C. slab roof of the conventional kitchen with a storage tank, to provide about 0.2 m^3 of hot water of 50-65°C for various domestic purposes. During harmattan and rainy period, solar energy is to be supplemented by fuelwood for water heating.

J4.2

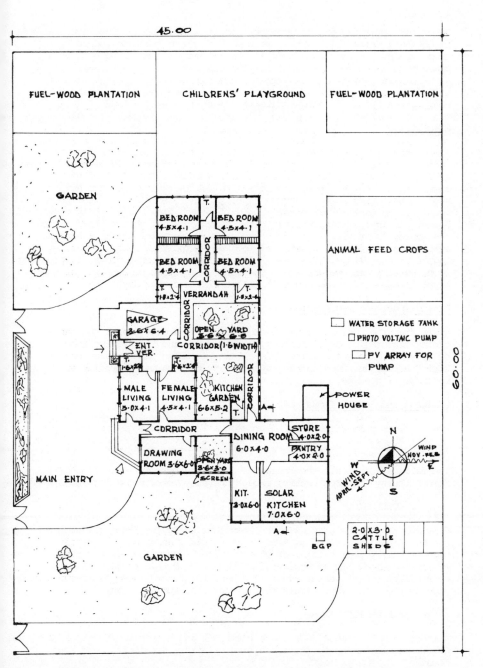

FIG.1. GROUND FLOOR PLAN OF ENERGY SELF-SUFFICIENT HOUSE

J4.3

332   S. C. Bajpai and A. T. Sulaiman

Water supply

Water for domestic, animal, poultry and irrigation of gardens, etc. has to
be obtained using a 198 Watt peak power solar photovoltaic module for the
pump system. Photovoltaic modules for the pump are to be mounted at a
place such as to avoid shadowing. The pump would be able to supply about
10 m^3 of water every day at average sunshine level. A storage tank of
6.0 m^3 capacity would be provided about 4 m above the ground. The daily
consumption of water in the house is expected to be 3 - 4 m^3.

Electricity

All electricity requirements of the house are to be met by photovoltaic
energy conversion using monocrystalline silicon solar cells. About 57.5 m^2
area of photovoltaic modules are to be installed on the roof of dining
room, store, pantry, power house and toilet including the roof hangings. 30
storage batteries each of 100 Ah capacity are to be placed in the power
house with a battery protector to provide hitch free power supply to the
house. An inverter is to convert the d.c. output of photovoltaic array
and storage batteries into 220 V a.c. voltage to power normal electrical
systems in the house. Electrical load would include lights, security
lights, fans, fridge, TV, video , carpet cleaner, kitchen utensils, pressing
iron, passive air coolers, etc. The total electrical consumption of the
house is expected to be about 15 kWh at 220V a.c. per day.

Living space cooling/Heating

Apart from the cooling to be obtained from the architectural and lands-
caping measures, each of bedroom, living rooms and the drawing room are to
be provided with evaporative type of cooling using passive air coolers.
During harmattan, wood burning systems are to be used to provide space
heating, if need be.

Fuelwood and Animal Feeds

The fuelwood plantations are to provide all fuelwood requirements in the
house. Animal feed crops are expected to supply adequate amounts of feed
for about 15 domestic animals and 50 poultry birds to be kept in the house.

The fabrication of all the renewable energy systems is to be done at the
Sokoto Energy Research Centre, except the photovoltaic modules and systems.

CONCLUSION

Previous proto-type studies have shown that energy self-sufficient houses
can be a reality. Based on similar principles, an actual moderately big
size of energy self-sufficient house has to be constructed, occupied and
closely studied. Experiences, if positive from this project with the
cost effectiveness factor may form a base for future energy self-sufficient
housing programmes, particularly for rural and remote areas.

ACKNOWLEDGEMENTS

Authors wish to acknowledge the suggestions and help given by Mr. M.M.Sharma,
Chief Architect, Sokoto State Ministry of Works, Transport, Land and
Housing, in this project.

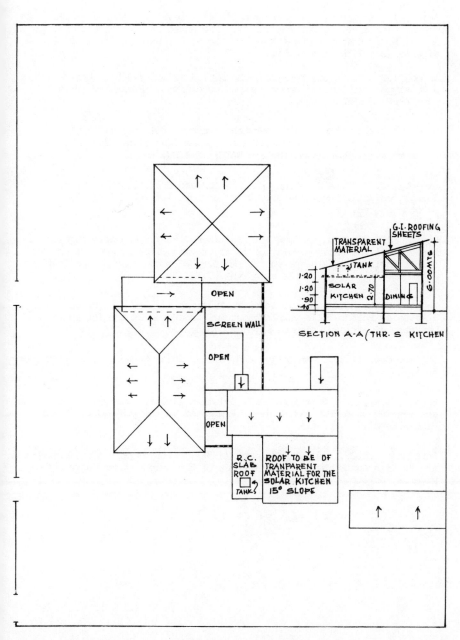

FIG.2. ROOF PLAN OF ENERGY SELF-SUFFICIENT HOUSE

REFERENCES

Ahmed, I., M. Mauzu, A.T. Sulaiman (1984). Development of a simple cooker with in-build Hot Box Effect. Presented at the 4th National Annual Solar Energy Forum (NASEF) of the Solar Energy Society of Nigeria (SESN), Birnin Kebbi.

Ahmed, I., M.M. Garba, M. Mauzu, and A.T. Sulaiman (1985). Evaluation of a Box-type solar cooker under various conditions. Presented at the 5th NASEF of SESN, Enugu.

Ahmed, I., S.C. Bajpai, B.G. Danshehu and A.T. Sulaiman (1985). Fabrication and Performance Analysis of Some Traditional and Improved Woodstoves. Presented at the 5th NASEF of SESN, Enugu.

Akujobi, C.O. (1984). Design, Fabrication and Characterisation of Integrated Solar Water Heater. B.Sc. Honours Thesis, University of Sokoto, Sokoto.

Atiku, A.T., S.C. Bajpai and A.T. Sulaiman (1985). Characterization of Monocrystalline Silicon solar cells in varying Climatic conditions in Sokoto Environment. Prepared for the UK-ISES conference an Applications of Photovoltaics, Newcastle upon Tyne, U.K.

Bajpai, S.C., A.T. Rostocki and A.T. Sulaiman (1985). Performance Study of a Solar Photovoltaic Pumping System. Presented at the International Workshop of African Union of Physics, Ibadan.

Bajpai, S.C. (1985). An Active and Passive Solar House. J. Ambient Energy 6, 1, 25 - 30.

Bugaje, I.M., A.B. Saraki and A.T. Sulaiman (1984). Design, Construction and Testing of a Room Air Cooler. Presented at the 4th NASEF of SESN, Birnin Kebbi.

Fernando, C.E.C., and S.M. Dangoggo (1984). Studies on Biogas. Presented at the 25th Annual National Conference of the Science Association of Nigeria, Sokoto.

List of Participants

AARSEN, F G van den
BECE - Biomass Energy Consultants
P O Box 498
7600 Al Almelo
THE NETHERLANDS

ABELES, Dr T P
Inst.Le Associates
3702 E.Lake
Minneapolis MP 55405
USA

AGNEW, P
Portree Cottage
Alme Ave
Aberfeldy
Perthshire
SCOTLAND

AL-AYFARI H H
University of Birmingham
Mechanical Eng.Dept.
Birmingham B15 2TT
ENGLAND

ANDREWS, Dr D
Energy Research Group
The Open University
Walton Hall
Milton Keynes,MK76AA
ENGLAND

ARAFA, Prof. Salah,
Science Dept.
The American University in
Cairo,
113 Kasr El-Aini St.
Cairo, EGYPT

BANNISTER, S
Napier College
Colinton Rd.
Edinburgh EH10 5DT
SCOTLAND

BAJPAI, S C,
Sokoto Energy Research Centre
University of Sokoto,
PMB 2346, Sokoto,
NIGERIA

BARNES, P M L
SIPC PL/16
SHELL CENTRE
London SE1
ENGLAND

BARP, B
Sulzer Bros.Ltd.
Research Dept.
CH-8401 Winterthur
SWITZERLAND

BASON, F
Linabakken 13
DK-8600 Silkeborg
DENMARK

BEEVERS, DR C L
SHELL
Non Traditional Business
Shell Centre
London SE1
ENGLAND

BENEVOLO, G
AGIP NUCLEARE
Piazza Ludovico Cerva 7
00143 Roma
ITALY

BERNARDO, Rafael O
Manager, Engineering Design
PNOC Energy Development Corp.
PNOC Complex,
Merritt Road
Fort Bonifacio
Metro Manila
PHILIPPINES

BEYER, H G
Universitat Oldenburg
FB Physik
Postfach 2503
D2900 Oldenburg
WEST GERMANY

BIET, B
FAO
Regional Office for Europe
Via delle terme di Caracalla
00100 Rome
ITALY

335

BODRIA, Dr L
FAO
University of Milan
Inst.for Agric.Eng.
Via Celoria 2,
20133 Milan
ITALY

BONKE, Knut,
Kvaerner Brug A/S
Postboks 3610 GB
Oslo, 1,
NORWAY

BOSSANYI, Dr E,
Rutherford Appleton Lab.
Chilton, Didcot,
Oxon OX11 OQX
ENGLAND

BRETT, E
Clachanduish Farm House
Balvicar,
Oban, Argyll
SCOTLAND

BRICKENDEN, D
Herbst Servotherm Ltd
Kilpoole Hill
Wicklow
IRELAND

BRITTEN, John R,
BRECSU (Building
Research Establishment)
Garston, Watford,WD2 7JR
ENGLAND

BROOKS, P
Alcan International
Southam Road
Banbury
Oxon
ENGLAND

BROWN DOUGLAS,G R,
Mansfield, Chapelton
By Strathaven
Lanarkshire ML10 6SG
SCOTLAND

CAIRNS, R J
Systematic Micro Ltd.
Index House
Ascot
Berks SL5 7EU
ENGLAND

CAMERON, D A
Countryside Commission for Scotland
Battleby, Redgorton
Perth
SCOTLAND

CARNIE, C
18 Woodside Crescent
Glasgow G3 7UU
SCOTLAND

CASTELLI, Prof.G
FAO
University of Milan
Inst.of Agric.Eng.
Via Celoria 2,
20133 Milan
ITALY

CAVE, Dr P R,
Dept.Mechanical Eng.
Plymouth Polytechnic
Drake Circus
Plymouth PL4 8AA
ENGLAND

CLARKE, T
Farm Gas Ltd
Industrial Estate
Bishop's Castle
Shropshire SY9 5AQ
ENGLAND

CRAWFORD, M
7A Claremont Cres.
Edinburgh EH7 4HX
SCOTLAND

DICKSON, F
Ogilvie Dickson Ltd.
Wylies Brae
Galashiels
SCOTLAND

DUFFIE, J A
University of Wisconsin-
Madison,
College of Eng.
Solar Energy Laboratory
1500 Johnson Drive,Madison
Wisconsin 53706
U S A

EJEBE, DR G C
Electrical Engineering Dept.
Faculty of Engineering
University of Nigeria
Nsukka
NIGERIA

ELGAMMAL, DR A
Mech.Eng.Dept.
Faculty of Engineering
Alexandria University
Alexandria,
EGYPT

ELLIOT, G
N E L
East Kilbride
Glasgow G75
SCOTLAND

EVANS, F
Physics Dept.
University of St Andrews
St Andrews
Fife KY16 9SS
SCOTLAND

EVANS, J H
Holec Ltd.
Burton House
1-13 High Street,
Leatherhead
Surrey KT22 BAA
ENGLAND

EVEMY, P
The Warmer Campaign
83 Mount Ephraim
Tunbridge Wells
Kent TN4 8BS
ENGLAND

FALCHETTA, M
ENEA-CASACCIA Dept.Fare
Sacco No.99 0060 A M~Gneria
Rome
ITALY

FARRELL, R M
Altamont Energy Corp.
1330 Lincoln Avenue
Suite 201
San Rafael
California 94901
USA

FAWKES, I & J
Marlec Eng.Co.Ltd.
Unit 5 Pillins Rd
Oakham
Rutland LE15 6QF
ENGLAND

FELSENSTEIN, G
Agric.Research Organ.
The Volcani Centre
Inst.of Agric.Eng
P O Box 6
Bet Dagan 50-250
ISRAEL

FERON, P
University of Reading
Dept.of Engineering
PO Box 225
Reading RG6 2AY
ENGLAND

FINDLAY, I
Energy Services Ltd.
8A Somerset Place
Glasgow
SCOTLAND

FORD, J A H
North of Scotland Hydro
Elecric Board
Dingwall
Ross-shire
SCOTLAND

FREDRIKSEN, A E
University of Trondeim
Institute of Physics
N-7055 Dragvull
NORWAY

FULTON, A M
Arch.Services Dept.
H R C Regional Bldgs.
Glenurquhart Rd.
Inverness
SCOTLAND

GARDNER, P
Rutherford Appleton Lab.
Chilton, Didcot
OX11 OQX
ENGLAND

GARSIDE, John,
Cranfield Institute of
Technology, Cranfield
Bedford MK43 OAL
ENGLAND

GEORGE, R
University of Bath
School of Biological Sciences
Bath BA2 7AY
ENGLAND

GERMAN, J.C.
Assistant Radio Engineer,
Northern Lighthouse Board
84 George Street
Edinburgh   EH2   3DA
SCOTLAND

GRANT, Dr A
Department of Thermodynamics
University of Strathclyde
Glasgow
SCOTLAND

HALLIDAY, J
Rutherford Appleton Lab.
Chilton, Didcot
Oxon, OX11 OQX
ENGLAND

HANNA, W R
Energy Division
National Board for Sci.& Tech.
Shelbourne House
Shelbourne Road
Dublin 4
EIRE

HARRISON,J D L
ODA, room E323
Eland House
Stag Place
London SW1E 5DH
ENGLAND

HART, Dr D
University of East Anglia
Environmental Sciences
Norwich Nr4 7TJ
ENGLAND

HARWOOD, Mr Mike
Technological Economics
Reseach Unit
University of Stirling
Stirling FK9 4LA
SCOTLAND

HAYES, Fergus
Bonnenberg & Drescher
9 Northumberland Rd.
Dublin, 4
EIRE

HERBST, M
Herbst Peat & Energy Ltd.
Head Office Kilpoole Hill
Wicklow
IRELAND

HOOPER, P N,
James Williamson
& Partners,
231 St Vincent Street
Glasgow G2 5QZ
SCOTLAND

HOPFE, Harold H,
Technical Advisor
U S Wave Energy Inc.
65 Pioneer Drive
Longmeadow,
Mass.01106
U S A

HOUNAM,I
Energy Studies Unit
University of Strathclyde
Glasgow G1
SCOTLAND

JAGADEESH, Dr A.
2/210 Nawabpet,
Nellore - 524 002
A.P. INDIA

JAMES, W
Solar Energy Research Inst.
of Western Australia
Sereway GPO box R1283
Perth,
W AUSTRALIA 6001

JANSSON, I
Lbt. PO Box 627
S-22006 Lund
SWEDEN

JANUARY, M
Blaenau Ucha Farm
Ffordd-Y-Blaenau
Treuddyn
Nr Mold
Clwyd Ch7 4NS
WALES

JELINKOVA, H
Research Inst.of AG Eng.
Ksanan 50
16307 Prague 6, Repy.
CZECHOSLOVAKIA

JENJE, Andrew A M,
Nedalo Bv
Uithoorn 1422 aJ
Industrieweg 4
HOLLAND

JONES,I
Dept.of Civil Engineering
University of Salford
Lancashire
ENGLAND

JORGENSEN, Dr Kaj
Physics Laboratory 111
The Technical University of
Denmark
Building 309 C
DK 2800 Lyngby
DENMARK

JORGENSON H F
Energy Studies Unit
University of Strathclyde
Glasgow G1
SCOTLAND

KERR, A
Dower House
Melbourne
Derby DE7 1EN
ENGLAND

KIRBY, Tim
N.C.A.T.
Machynlleth
Powys
WALES

KJAERBOE   Mr P
Hagagrand 1
S113 27
Stockholm
SWEDEN

KULCSAR, P
Nat.Inst.of Agric.Eng.
Tessedik Samuel u4,
2101 GODOLLO
HUNGARY

KYRITSIS, Prof.S
Dept.of Agricultural Bldgs.
75 Iera Ados
Botanikos
Athens
GREECE

LEESON, M R
University of Technology
Mechanical Engineering
Loughborough
Leics LE11 3TH
ENGLAND

LESMO, Dr Renato
v J Kennedy
16 - 20097 S Donato,
Mil. 1
ITALY

LEWIS, Dr C W
Energy Studies Unit
UNiversity of Strathclyde
Glasgow
SCOTLAND

LEWIS, Mr
Applied Energy Systems
1 Whippendel Road
Watford, Herts
ENGLAND

LOW, J
General Techn.Systems Ltd.
Forge House
20 Market Place
Brentford
Middlesex TW8 8EQ
ENGLAND

LUNDSBERG, M
Alternegy Aps
Mullerup Havn
Nordvej 4-6
4200 Slagelse
DENMARK

LUTHER, PROF Dr J
Fachbereich 8
Physik
Universitat Oldenburg
Postfach 2503
D-2503 Oldenburg
FED REP OF GERMANY

McCALLUM, W
School of Architecture
College of Art
Perth Road
Dundee DD1 4HT
SCOTLAND

McCARTHY, Sean
National Microelectronics
Research Centre
Lee Maltings, Prospect Row,
Cork,
REPUBLIC OF IRELAND

McGREGOR, Mr Kerr,
Chem.Eng.Dept.
Napier College
Colinton Road,
Edinburgh
SCOTLAND

MacLEAN, Dr C
Thurso Technical college
Thurso
SCOTLAND

MacLeod, I
Western Isles Island Council
Planning & Dev. Dept.
Sandwick Road
Stornoway
Isle of Lewis
SCOTLAND

MALCOLM, T S
HIDB, Bridge House
Bridge Street
Inverness IV1 1QR
SCOTLAND

List of Participants

339

McINNES, G
AFME
Service de l'Action Internationale
27 Rue Louis - Vicat
75015 Paris
FRANCE

MATHERS, M
SIAE, Penicuik
Midlothian
SCOTLAND

MELVILLE, I
Industry Dept.for Scotland
Energy Div.
New St.Andrews House,
Edinburgh EH1 3TA
SCOTLAND

MORBETY, T
Estacao Nacional de
Technologia does Produtos
Agrarios - INIA
Quinta do Marques
2780 Oieras
PORTUGAL

MORETON G B
F Banford & Co Ltd.
Ajax Works
Whitehill
Stockport
Cheshire SK4 1NT
ENGLAND

MORRIS, R M
Scottish Development Agency
Planning & Projects Dir.
120 Bothwell Street
Glasgow G2 7JP
SCOTLAND

NEILSEN, H T
Riso National Lab.
Frederiksborgvej 399
DK 4000 Roskilde
DENMARK

NGOKA, Mr Nelson I,
Head
Building Energy Conservation
Research Group (BECRG)
University of Ife
P O Box 1035
Ile-Ife
NIGERIA

ODUKWE, PROF. A O
National Centre for
Energy Research &
Development
University of Nigeria
Nsukka
NIGERIA

O'BROLCHAIN, F
Bonnenberg & Drescher
9 Northumberland Rd
Dublin 4
EIRE

OKSANEN, Dr Erkki,
Box 28
SF-00211 Helsinki,
FINLAND

OST, S
VUJ AB P O Box 519,
S-162 15 Vallingby
SWEDEN

OZMERZI, Aziz
Ankera
Universiteoi Zuraot
Fakiltesi Ankera
TURKEY

PAGE, Dr D I,
Energy Technology Support Unit
Harwell Laboratories,
Didcot
OX11 ORA
ENGLAND

PEATFIELD, A.M
Coventry (Lanchester)Poly
Priory Street,
Coventry CV1 SFB
ENGLAND

PELLIZZI,Prof.
University of Milan
Inst. of Agric.Eng.
Via Celoria 2
20133 Milan
ITALY

PETERSON,C M
12 Stansby Park
The Reddings
Cheltenham
Gloc.GL51 6RS
ENGLAND

PIRAZZI, L
S O Anguillarese No.301
ENEA CRE CASACCIA 0060
S. MARIA DI GALERIA
Rome
ITALY

POWLES, DR Simon
C/o Sir Robert MacAlpine
1 St Alban`s Road,
Hemel Hempstead
Herts. HP2 4TA
ENGLAND

PRINS, J P   UNDP
CHIEF, Unit for Europe
One United Nations Plaza
New York, NY 10017
U S A

PUIG, J
Dept. of Geography
Autonomous University of
Barcelona
Bellaterra
Catalunya
SPAIN

RASMUSSEN, MR H C
Energy Research Laboratory
Niels Bohrs Alle 25
DK 5230 Odense
DENMARK

REILLY, J
Nord Tank A/S, Nyballe VEJ 8
DK 8444 Balle
DENMARK

RESTIVO, Dr A
Universidade do Porto
Dept.Eng. Mec.
Rua dos Bragas
4000 Porto
PORTUGAL

RIVA, G
University of Milan
Inst.of Agric.Eng.
Via Celoria 2
20133 Milan
ITALY

ROBERTSON, Elliot
Tigh Geal
North Connel
Oban
PA 37 1QZ

ROBINSON, Mr Paul,
Dept. of
Electrical & Electronic
Engineering
Plymouth Polytechnic
Drake Circus
Plymouth, Devon
ENGLAND

RODGER, Mr D Elliot,
Energy Mines & Resources
Canada
Energy Program,
Ottawa, Ont.K1A OE4
CANADA

RUDOLPH, J
Eastern Energy
The Elms
Little Carlton
South Lincs LN11 8HN
ENGLAND

SALTER, S H,
Dept.of Mech.Eng.
Univesity of Edinburgh
The King's Buildings
Edinburgh EH9 3JL
SCOTLAND

SCOTT, Mr Richard,
Institute of Terrestrial
Ecology, Merlewood Research
Station, Grange-over-Sands,
Cumbria LA11 6JU
ENGLAND

SHULTZ, Prof.E B,
School of Engineering
&Applied Sciences
Dept.of Engineering & Policy
Washington University
Campus Box 1106,St Louis,
Missouri 63130,
U S A

SMITH, P
N.E.W. Ltd. Tanners Yard
Gilesgate,
Hexham,
Northumberland
ENGLAND

SODERHOLM, Dr.Leo,
Rm 213 Davidson Hall
Iowa State University
Ames Iowa 50010
U S A

SOMERVILLE, D
Ross & Cromarty District Council
County Buildings
Dingwall
SCOTLAND

STAMPACH, Svatopluk
Federal Ministry of Agriculture
Tesnov 17
11006 Prague 1
CZECHOSLOVAKIA

STANCAMPIANO, Maurizio
Soc Conphoebus
Via G Leopardi 6
95127 Catania
ITALY

STEINBERGER, Mr R
Fachbereich 8/Physik
Universitat Oldensburg
Carl-von-Ossietzky-Str.
D-2900 Oldenburg
WEST GERMANY

SULAIMAN, A T
Sokoto Energy Research Centre
University of Sokoto
PMB 2346 Sokoto
NIGERIA

TAYLOR, J
N.E.W. Ltd., Tanners Yard,
Gilesgate,
Hexham
Northumberland
ENGLAND

TAYLOR, Dr R
Bld.156, ETSU
AERE, Harwell
Oxon, Ox11 ORA
ENGLAND

TEACI, Dr D
Academy for Agric.&
Forestry Sciences
Bulevard Marasti 61
Bucharest
ROMANIA

THOM, S
Melbourne University
Dept.of Mech.Eng.
Graffan St
Parkville 3052
AUSTRALIA

THOMAS, Antony
Instrumentation & Services
Unit,
Indian Institute of Science
Bangalore 560 012
INDIA

TOMTER, N
Scottish Peat & Land Dev.
18 Strathearn Rd.
Edinburgh EH9 2AE
SCOTLAND

TUDOROVIC, Dr M
University of Belgrade
Institute of Agricultural Engineering
Nemanjina 6
11080 Zemun
YUGOSLAVIA

TWIDELL Dr J W
Energy Studies Unit
University of Strathclyde
Glasgow G1
SCOTLAND

TYMINSKY, Prof.J
Institute for Building,
Mechanization & Electrification
in Agriculture
ul. Rakoviecka 32
02-532 Warsaw
POLAND

VENNER, G
51 Hermitage Rd.
Plymouth PL3 4RX
ENGLAND

VIRGILI, Alberto
Soc.Conphoebus
Via G Leopardi 60
95127 Catania
ITALY

WALKER, J A
Wallace Whittle & Partners
6/7 Newton Terr.
Glasgow G3 7P3
SCOTLAND

WASPODO, Priyo
3 Leven Street
Glasgow G41 2JB
SCOTLAND

WELSCHEN, J
R & S   Renewable Energy
Systems
Croy 49 5653 LC Eindhoven
NETHERLANDS

WESTGATE, Mr Michael,
Peddocks Island Trust
45 School Street, Boston,
Massachusetts 02108
U S A

WHELDON, Dr Anne
Physics Dept.
University of Swaziland
Kwalusani
SWAZILAND

WHITE, Dr Peter,
Energy Group
Coventry (Lanchester)
Polytechnic
Priory Street
Coventry CV1 5FB
ENGLAND

WHITE, R J
Easterachloa
Duneaves Road
Aberfeldy,
Perthshire PH15 2LS
SCOTLAND

WILLIAMS
Ross & Cromarty District
Council
County Buildings
Dingwall
SCOTLAND

WILSON, Dr D
Cavendish Laboratory
Madingley Road,
Cambridge CB3 OH
ENGLAND

WILSON-GOUGH, A
Beenleigh Manor
Harbertonford
Totnes
Devon
ENGLAND

XIANGFAN, Gao
The Guangzhou Institute of
Energy Conversion
Chinese Academy of Sciences
P O Box 1254
Guangzhou
CHINA

ZHU, G
Institute of Energy Research
Rural Energy & New Energy Dept.
Building 91
Bei Sha Tan
Beijing
CHINA

Author-Paper Index

(See List of Participants for addresses and institutions)

AUTHOR	PAPER	PAGE
Arafa, S	A3	15
Atiku, A T	A6	29
Bajpai, S C	A6, J4	29,329
Bergman, D A	C6	115
Bernardo, R O	H1	259
Bonke, K	G3	249
Britten, J R	J1	311
Buckland, M P	C6	115
Callaghan T Y	C6	115
Carnie, C G	E2	151
Cave, P R	E3	159
Clarke, J M	C4	101
Duckers, L J	E5	171
Duffie, J	I1	275
Evans, E M	E3	159
Fernando, C E C	A6	29
Fredriksen, A E	E6	179
Gabler, H	J2	317
Gardner, P	F1	185
Garside, J	A7	35
Gerdes, G J	J2	317
German, J C	F5	213
Gold, B D	C2	87
Grant, A D	G2	243
Guillen, H Y	H1	259
Guven, H M	I2	281
Halliday, J	F1	185
Harwood, M	J3	323
Herbst, M	C3	93
Hooper, P N	E1	143
Hopfe, H H	G2	243
James, W	I5	301
Jaques, J K	J3	323
Jenje, A M	C5	107
Jones, I	E2	151

Subject Index

345